HARRAP'S

Maths

MINI DICTIONARY

Consultant Editor
Nigel Andrews

HARRAP
Edinburgh New York

First published in Great Britain 1993 by

Harrap Books Ltd.
Chambers Publishers
43–45 Annandale Street
EDINBURGH EH7 4AZ

ISBN 0-245-60472-3

Typeset by J W Arrowsmith Ltd., Bristol
Printed in England by Clays Ltd, St Ives plc

Preface

This Mini Dictionary is intended for use by two principal groups of readers: students following mathematics courses to GCSE level; and general readers seeking concise definitions of the more basic terms and concepts in mathematics. Worked examples showing how particular prodedures are applied in detail are provided in many instances. There is also a separate Appendix (II) on the use of pocket electronic calculators, and a number of the worked examples are written for solution with a calculator. Appendix I lists conversions and other useful formulae. Appendix III gives brief biographical information on mathematicians whose work has a particular bearing on terms and concepts which appear in the Mini Dictionary.

The initial term in an entry appears in **bold**; related terms which crop up in an entry with their own entries elsewhere in the Mini Dictionary are also shown in bold. Where the reader is unsure of the meaning of an abbreviation or symbol, reference should be made to the separate lists of abbreviations and symbols at the beginning of the Mini Dictionary.

GCSE and the National Curriculum

In order to help students using this book as a study aid, entries which are particularly relevant to National Curriculum Attainment Targets in Mathematics are indicated by ■ at the beginning of the entry.
Such entries also include a precise indication of the relevant Attainment Target in the form of a bracketed sequence of numbers and letters, as in [5/10/b]. The first letter is the Attainment Target number; the second is the Attainment Level, and the letter at the end is the relevant section of the Attainment Level.

ABBREVIATIONS

A list of the abbreviations, mathematical and non-mathematical, used in the Mini Dictionary follows. Unless otherwise stated, abbreviations for a single unit also apply to plural units, e.g. hr = hour; hours.

A	ampere, area
acos	arc cosine
ADC	analog/digital converter
aln	antilogarithm
amp	amplitude
AP	arithmetic progression
APR	annual percentage rate
arg	argument
asin	arc sine
atan	arc tangent
ATM	automatic teller machine
AU	astronomical unit
BIN	binary
C	Celsius
c.	approximately (*circa*)
cc	cubic centimetre
cd	candela
c.g.s.	centimetre, gram, second
cm	centimetre
cm^2	square centimetre
cos	cosine
cosec	cosecant
cosech	hyperbolic cosecant
cot	cotangent
coth	hyperbolic cotangent

cwt	hundredweight
DEC	decimal
DTP	desktop publishing
E	East
exp	exponent
F	Fahrenheit
fl.	lived (*floreat*)
fl oz	fluid ounce
f.p.s.	foot, pound, second
ft	foot; feet
ft^2	square foot
ft^3	cubic foot
g	gram
gal	gallon
GP	geometric progression
ha	hectare
HCF	highest common factor
HEX	hexadecimal
hr	hour
hyp	hyperbolic function; hypotenuse
I	interest
in	inch
in^2	square inch
in^3	cubic inch
K	kelvin
kg	kilogram
km	kilometre
km^2	square kilometre
km^3	cubic kilometre
km/h	kilometres per hour
kn	knots
l	litre
lb	pound
LCD	lowest common denominator

LCM	lowest common multiple
ln	Napierian logarithm
log	common logarithm
M	mile
m	metre; mean
m^2	square metre
m^3	cubic metre
max	maximum
$mile^2$	square mile
$mile^3$	cubic mile
min	minimum; minute
ml	millilitre
mm	millimetre
mod	modulus
mol	mole
mph	miles per hour
m/s	metres per second
N	North
OCT	octal
oz	ounce
P	principal
PC	personal computer
p.a.	per annum
pr	probability
pt	pint
QED	*quod erat demonstrandum*
QEF	*quod erat faciendum*
qt	quart
R	rate
r	radius
rad	radian
rms	root mean square
ROM	read-only memory
S	set; sum; South

s	sccond
sd	standard deviation
sec	secant
sech	hyperbolic secant
sf	significant figure
shm	simple harmonic motion
sin	sine
sinh	hyperbolic sine
SI	Système Internationale
sq	square
sr	steradian
tan	tangent
tanh	hyperbolic tangent
V	volume
VAT	value-added tax
W	West
yd	yard
yd^2	square yard
yd^3	cubic yard

SYMBOLS

A list of the mathematical symbols used in this Mini Dictionary follows. For convenience, this is divided into symbols which are letters in the Roman alphabet (the alphabet used in this Mini Dictionary), symbols which are letters in the Greek alphabet, and symbols which are not letters in any alphabet currently in widespread use.

Roman letters

A	duodecimal 10
$\vec{\mathrm{A}}$	vector (here vector A)
A	hexadecimal 10
a	acceleration
a	atto- ($\times 10^{18}$)
$\frac{a}{b}$	fraction
a/b	a divided by b
$a:b$	a proportional to b
B	duodecimal 11
B	hexadecimal 11
C	hexadecimal 12
c	centi- ($\times 10^{-2}$)
D	hexadecimal 13; Roman numeral 500
d	diameter of a circle
d	deci- ($\times 10^{-1}$)
da	deca- ($\times 10$)
E	duodecimal 11
E	hexadecimal 14; exa- ($\times 10^{18}$)
e	2·71828 . . . ; eccentricity (coefficient of restitution)

e^x	Napierian antilogarithm; exponential function
F	hexadecimal 15
f	femto- ($\times 10^{-15}$)
f'	derivative
$f(x)$	function x
G	giga- ($\times 10^9$)
h	height (of a geometric figure)
h	hecto- ($\times 10^2$)
I	moment of inertia; Roman numeral 1
i	$\sqrt{-1}$ (see **complex number**)
k	kilo- ($\times 10^3$)
L	Roman numeral 50
l	length (of a geometric figure)
M	mega- ($\times 10^6$); Roman numeral 10^3
m	mass
m	milli- ($\times 10^{-3}$)
$\mathrm{m\ s^{-1}}$	metres per second
N	newton
n	nano- ($\times 10^{-9}$)
n	any number
n^x	any number to a given power
n_x	any number to a given base
P	peta- ($\times 10^{15}$)
p	pico- ($\times 10^{-12}$)
s	distance
T	duodecimal 10
T	tera- ($\times 10^{12}$)
V	Roman numeral 5
v	vector
v	velocity
X	Roman numeral 10
x	unknown; horizontal Cartesian axis
$\frac{1}{x}$	reciprocal

x^{-1}	reciprocal
$\bar{x}$	arithmetic mean
x^y	number to a given power
y	unknown; vertical Cartesian axis

Greek letters

α	alpha
β	beta
μ	mu; micro- ($\times 10^{-6}$)
Ω	omega (capital letter)
ω	omega (small letter)
π	pi; 3·14159 . . .
Σ	sigma (capital letter); sum
σ	sigma (small letter); standard deviation
θ	theta; geometric angle of unknown measure

Non-letter symbols

$+$	add; positive
$-$	subtract; negative
$\times$	multiply
$\cdot$	multiply
$\div$	divide
$/$	divide
$:$	ratio
$^\circ$	degree
$<$	less than
$\leq$	less than or equal to
$>$	greater than
$\geq$	greater than or equal to
$=$	equal to
$\neq$	not equal to
$\approx$	approximately equal to

$\equiv$	congruent to
$\propto$	proportional to
$\pm$	plus or minus
$\angle$	angle
$(\ldots)$	rounded brackets
$\{\ldots\}$	set
$[\ldots]$	closed interval
$\sqrt{}$	square root
$\sqrt[3]{}$	cube root
!	factorial
∞	infinity
£	UK pound sterling
$'$	prime; minute of degree, foot, recurring decimal
$''$	double prime, second of degree, inches
$'''$	triple prime
.	decimal point
%	percent (divided by 100)
$\int$	integral
$\rightarrow$	maps to

set theory

$\cap$	intersection
$\cup$	union
$\varnothing$	empty set
$\{\ \}$	empty set
$\in$	element of
$\notin$	not element of
$\subset$	subset of

A

A Number 10 in the **duodecimal** number system.

A *1.* Number 10 in the **hexadecimal** number system. *2.* Abbreviation of **ampere**. *3.* Abbreviation of **area**.

a Symbol for **acceleration**.

a Symbol for **atto-**, $\times 10^{-18}$.

abacus Manual **calculator** in the form of a **rectangular** frame within which are fixed **parallel** wires or rods along which beads are moved in rows or columns. Calculations are performed by the user sliding the beads along the wires. In some abacuses the beads are moved horizontally (in rows), in others vertically (in columns). In almost all cases, however, the rows or columns represent **powers** in a number system. In a vertical **decimal system** abacus (as in Figure A1), for example, nine columns would represent (reading from right to left): units; $\times 10$; $\times 10^2$; $\times 10^3$; $\times 10^4$; $\times 10^5$; $\times 10^6$; $\times 10^7$; $\times 10^8$. A further aid to calculation can be added by splitting off the upper third of the columns and giving the beads in that upper third of the rectangle the value of $\times 5$ the beads in the lower two-thirds of the rectangle (also in Figure A1). Numbers are represented by sliding beads up or down towards the partition between the top third and lower two-thirds. In Figure A1, each column has seven beads: two in the upper third space, and five in the lower two-thirds space. The number shown in the Figure is 6,121,178 (i.e. read off how many of the seven beads in each column are arranged along the partition, not at the very top or bottom of the columns). Operations such as **addition** and **subtraction** are performed by moving beads towards and away from the partition. The abacus is one of the most ancient forms of calculator that we know about: different types were used by the ancient Egyptians, Greeks and Romans (i.e. over 2,000 years ago), and they were still in use in Europe as recently as 300 years ago when the use of the **logarithm** as an aid to calculation began to spread. Abacuses are still widely used in the modern world, for example in the Far East and the Middle East.

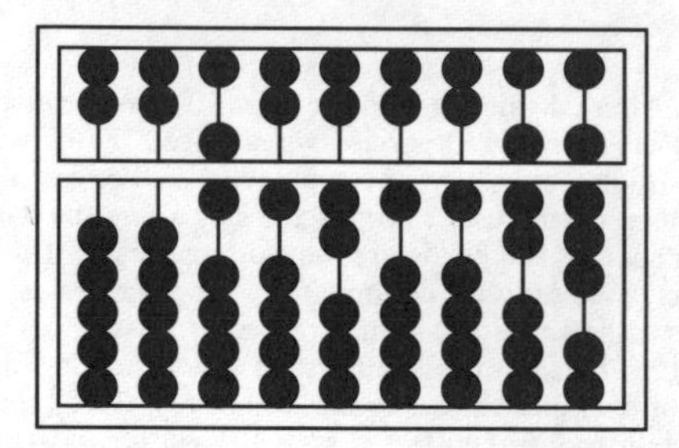

Fig. A1 *vertical abacus*

abscissa In coordinate geometry, the distance of a point from the y-axis in the direction of the x-axis. *See* **Cartesian coordinates**; **ordinate**.

absolute error **Difference** between an estimated value and the true value of a calculation. *Example*: if 80 is the estimated value, and 79·3 is the true value, the absolute error is 0·7. *See* **estimate**.

absolute value Numerical value of a **real number** irrespective of whether it is positive or negative or, if it is a **vector**, irrespective of its direction. Alternative name: **modulus** (or mod) x, where x is the number.

acceleration (a) Rate of change of **velocity**, normally expressed using the **SI units** m s^{-2} (metres per second). If velocity changes at a constant rate from an initial value of u to a final value of v in time t, acceleration a is given by:

$$a = \frac{(v - u)}{t}$$

Alternatively, if velocity is represented by v, acceleration is found by differentiating v with respect to time t:

$$\text{acceleration } a = \frac{dv}{dt}$$

See also **equations of motion**; **force.**

accuracy Measure of how close to, or distant from, a true value is an estimated value. *See* **estimate.**

acos Abbreviation for **arc-cosine.**

acre Unit of area on the **f.p.s.** system, equal to 4840 square **yards,** = 0·4046 **hectares.**

■ **acute** In **geometry**, describing an angle of less than 90° (as seen in Figure A2, where the angle BAC = 37°). [4/4/c]

ADC Abbreviation of **analog/digital converter.**

adder Part of a **computer** that adds digital signals (addend, augend and a carry digit) to produce the sum and a carry digit.

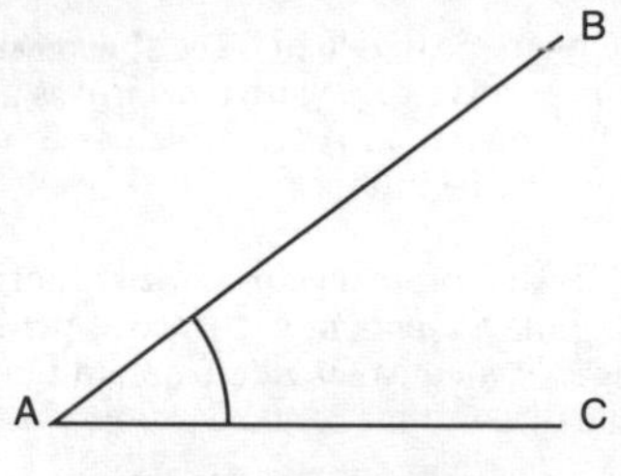

Fig. A2 *acute angle*

addition Basic operation of **arithmetic** for finding the **sum** of two numbers.

■ **addition of algebraic terms** Addition of numbers, both positive and negative, which can be applied using a **commutative** law i.e. $a+b=b+a$ and an **associative** law i.e. $a+(b+c)=(a+b)+c$. [2, 3 & 4]

■ **addition of binary numbers** Binary numbers use the base 2. They consist of the numbers 0 and 1. For example, decimal (denary) numbers 1, 2, 3 occur as 1, 10 and 11 in the binary system. [2/4/c] The rules of addition can be summarised as:

$$1+0=1$$
$$0+1=1$$
$$1+1=10$$

e.g.

$$\begin{array}{r} 10111 \\ +1101 \\ \hline =100100 \end{array}$$

Arithmetic operations in binary, and conversion of binary

numbers to other number bases, can be performed on most **scientific calculators**.

■ **addition of decimals** This takes place using exactly the same rules as for whole numbers. The numbers must be written with each decimal point directly under the one above. [2/4/1]
Example:

$$\begin{array}{r} 12{\cdot}43 \\ +2{\cdot}68 \\ +0{\cdot}006 \\ \hline =15{\cdot}116 \end{array}$$

■ **addition of fractions** Fractions can be added only if they are expressed with the same denominator, usually the **lowest common denominator** (LCD). [2/5/c] *Example*: the lowest common denominator of $\frac{1}{2}$, $\frac{1}{3}$ and $\frac{1}{4}$ is 12, and the fractions can be restated as $\frac{6}{12}$, $\frac{4}{12}$ and $\frac{3}{12}$. They can now be added:

$$\frac{6}{12}+\frac{4}{12}+\frac{3}{12}=\frac{13}{12}=1\frac{1}{12}$$

Arithmetic operations using fractions can be performed on most **scientific calculators** using the $a^{b/c}$ key.

■ **addition of matrices** If two matrices are of the same order they can be added. The order that should be followed is: the number of rows, then the number of columns. [4/10/d]
Example:

$$\begin{pmatrix} 1 & 3 & 4 & 5 \\ 2 & 6 & 8 & 4 \end{pmatrix}$$

is of order 2×4 because it has two rows and four columns. The number of rows always comes first.

Example: if $\begin{pmatrix} 2 & 3 \\ 4 & 5 \end{pmatrix}$ is added to $\begin{pmatrix} 6 & 7 \\ 8 & 9 \end{pmatrix}$

this bccomcs

$$\begin{pmatrix} 2+6 & 3+7 \\ 4+8 & 5+9 \end{pmatrix}$$

which equals

$$\begin{pmatrix} 8 & 10 \\ 12 & 14 \end{pmatrix}$$

■ **addition of vectors** A vector is a quantity that has size and direction, and is usually symbolized by a letter with an arrow above it, as in $\vec{A}$. Addition of vectors obeys the **associative** rule and the **commutative** rule (it does not matter in which order they are added). [4/8/d] [4/9/e] *Example*: in Figure A3, vector X + vector Y = the **resultant** Z, and $\vec{Y}+\vec{X}=\vec{Z}$.

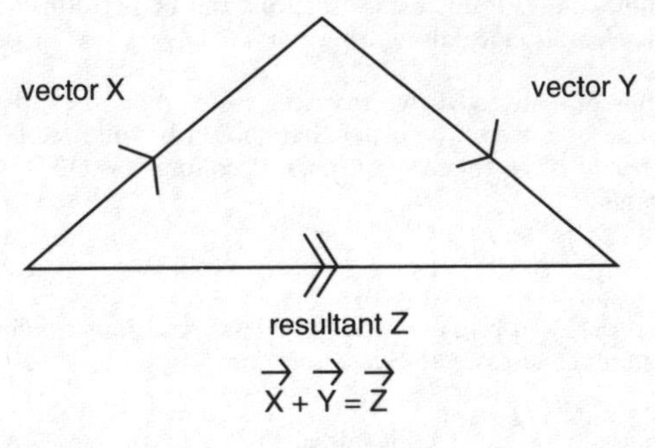

Fig. A3 *addition of vectors*

address In computing, *1.* identity of a location's position in a memory or store; or *2.* the specification of an operand's location.

■**adjacent** Lying next (or contiguous) to something. *Example*: side BC in the triangle ABC in Figure A4 is adjacent to angle a. The two sides CB and AB meet at the **vertex** B. [4/9/a]

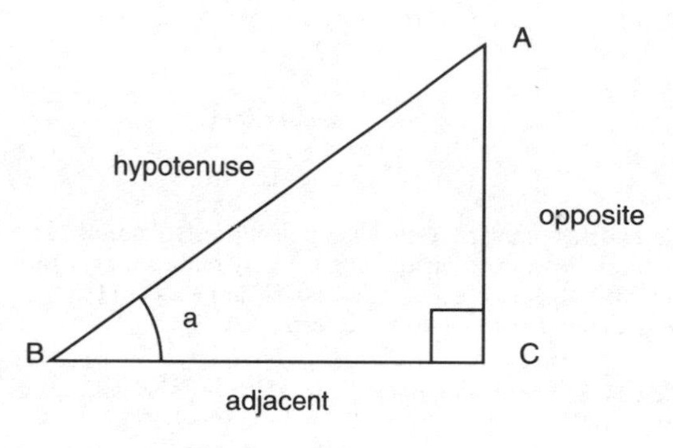

Fig. A4 *adjacent* (*side*)

In Figure A5, adjacent angles are angle a and angle b.

affine Type of **transformation** of a plane such as **enlargement**, **reflection** or **rotation** in which **parallel** lines remain parallel.

■**algebra** Branch of mathematics that deals with the general properties of numbers by means of abstract symbols. Typically a letter, such as x, is made to stand for an

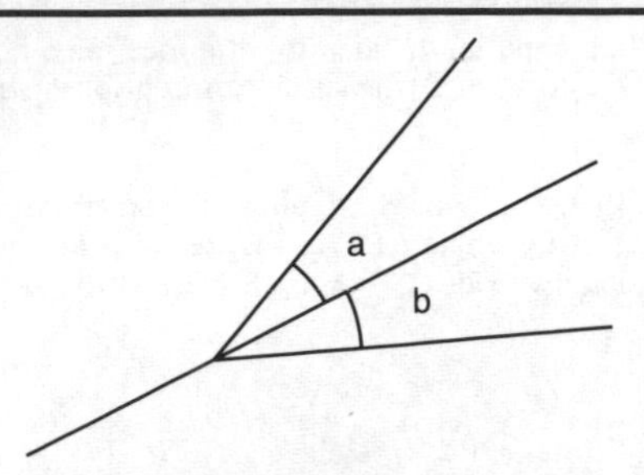

Fig. A5 *adjacent* (*angle*)

unknown quantity so that relationships involving it can be written down and manipulated. Many mathematical **formulae** involve algebraic expressions, which are also used in **coordinate** (or analytical) **geometry**. [AT3]

algebraic Describing operations which use the symbols of **algebra**.

ALGOL Acronym for the high-level computer programming language ALGOrithmic Language, used for manipulating mathematical and scientific data.

algorithm Operation or set of operations that are required to effect a particular calculation or to manipulate data in a certain way, usually to solve a specific problem. The term is commonly used in the context of computer programming.

aln Abbreviation for **antilogarithm**.

alphanumeric Describing characters or their codes that represent letters of the alphabet or numerals, particularly in computer applications.

■ **alternate angle** In geometry, one of a pair of angles on opposite sides of a line (a **transversal**) that cuts two other lines; thc alternate angles are those made where the transversal cuts the other lines. If the lines cut by the transversal are parallel, the alternate angles are equal. [4/4/c] [4/5/b] *Example*: in Figure A6, angles A, B, C, and D are the alternate angles. *See also* **corresponding angle**.

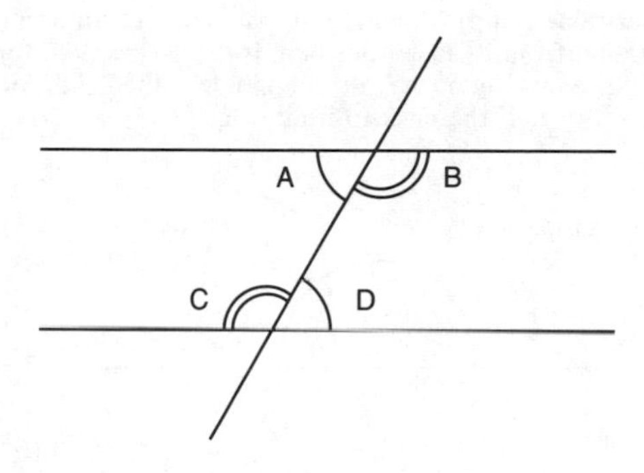

Fig. A6 *alternate angle*

altitude *1.* In a triangular **plane** or **solid** figure, the perpendicular distance from the **base** to the **vertex** opposite the base. In a solid, such as a **cone** or **pyramid**, the altitude is found by taking a plane section through the figure, and treating the section as a triangle. The **area** of a triangular

plane figure is found by the formula $\frac{1}{2} \times$ (altitude × base).
2. In a **rectangular** plane or solid, or in a **parallelogram**, the perpendicular height from the base to the opposite side of the rectangle or parallelogram. The area of a rectangle or parallelogram is found by the formula base × altitude. Alternative term: height.

amp Abbreviation for **amplitude**.

ampere (A) **SI unit** of electric current.

■ **amplitude** (amp) Maximum displacement of an oscillating motion from its mean position. [4/9/g] *Example*: for a **sine** wave, as in Figure A7, the amplitude is the height of the wave, or half the peak-to-peak value.

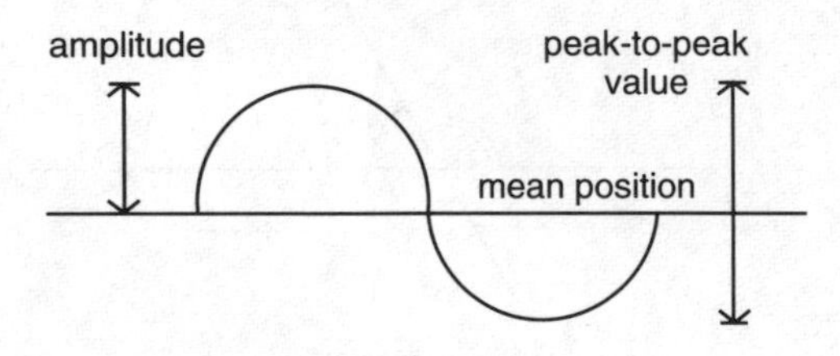

Fig. A7 *amplitude*

analog Method of recording values by means of a continuous wave or motion, as in the hands of a clock, or the rev counter and speedometer in a motor vehicle. Alternative term: analogue. *Compare* **digital**.

analog computer Computer that represents numerical values by continuously variable physical quantities (e.g. voltage, current). *Compare* **digital computer**.

analog/digital converter (ADC) Device that converts the output of an **analog computer** into digital signals for a **digital computer**.

analysis Branch of mathematics based on **convergence** and the limits of continuous **functions**. Examples are: **series**, **trigonometrical functions**, **calculus** and **logarithmic functions**.

analytical engine Mechanical calculating device invented by Charles Babbage (1792 – 1871). Babbage was unable to build the complete machine in his lifetime, but the London Science Museum has recently been able to do so. The engine is an important early form of the modern **digital computer**. *See* **calculator**; Appendix II.

analytical geometry *See* **coordinate geometry**.

AND gate Computer logic element that combines two binary input signals to produce one output signal according to particular rules. *Example*: in Figure A8, the circuit symbol

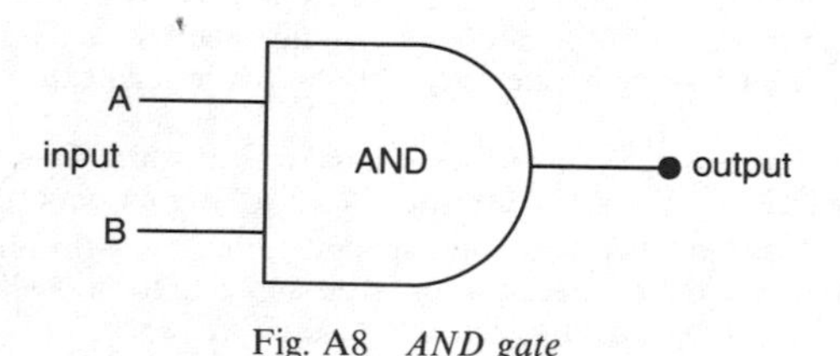

Fig. A8 *AND gate*

for an AND gate is shown. Output only occurs when there is an input to both A *and* B. Alternative name: AND element.

■**angle** Inclination of one line to another. Angles can be measured in **degrees** (there are 360 in a complete revolution), or **radians** (there are 2π in one revolution). [4/4/c]

angular *1.* Describing an **angle**. *2.* Measured in terms of angles.

angular distance In astronomy, the angle between two objects (usually stars) as perceived by an observer on Earth.

angular frequency Number of vibrations per unit time, multiplied by 2π, of an oscillating body. Usually expressed in radians per second.

angular momentum For a rotating object, the cross-product of a **vector** from a specified reference point to a particle and the particle's **linear momentum**.

annual Describing one year, equalling 12 calendar **months** =365 **days** in a normal year, 366 days in a **leap year**. A year usually begins on 1 January and ends on 31 December, but in financial accounts, an annual account can run from the beginning of any month in one year to the end of the previous month in the next. For example, the UK tax year runs from 1 April in one year to 31 March in the next.

annual percentage rate (APR) Annual rate at which **interest** is charged on a loan agreement. This is always an important consideration when borrowing money, for example through a **Consumer Credit Agreement**, or when applying for a **credit card** or purchasing direct on credit. *Example*: a person buying an item priced £300·00 on a credit agreement with an APR of 20% will have paid the equivalent of:

$$\frac{£300 \times 20}{100} + £300 = £360$$

at the end of a year if the full amount has not been paid in that period of time. *See also* **compound interest**; **simple interest**

annular eclipse Eclipse in which a thin ring of the source of light appears around the obscuring body.

annulus *1.* In geometry, a plane surface bounded by two concentric circles (i.e., the shape of a metal washer or a disc with a central hole in it). If the larger and smaller **radii** are R and r, the area of an annulus is $\pi(R^2 - r^2)$. *2.* In astronomy, a ring of light seen during an **annular eclipse**.

anticlockwise Describing **rotation** in a direction opposite to that normally followed by the hands of an **analog** clock or watch. *Compare* **clockwise**.

antilogarithm (aln) Number that is represented by a **logarithm**; *Example*: the antilogarithm of the **common logarithm** 0·3010 is 2, because the common logarithm of 2 (logarithm to the base 10) is 0·3010. The antilogarithm of the **Napierian logarithm** 0·6931 (logarithm to base e) is also 2. Antilogarithm functions for both common and Napierian logarithms can be found on most **scientific calculators**, and also on tables of common and Napierian antilogarithms.

AP Abbreviation for **arithmetic progression**.

apex Point on a **solid** or **plane** figure that is farthest from its base, for example the pointed top of a **cone** or the angle furthest from the base line of any regular triangle.

aphelion Point on the **orbit** of a planet (or comet) at which it is at its greatest distance from the Sun. *Compare* **perihelion**.

aposteriori probabilities *See* **Bayes's theorem**.

applications program Computer **program** written by the user for a specific purpose, for example record keeping or stock control.

■ **approximation** Near guess at a number. It is often used to find rough answers to computational problems. [2/5/g] *Example*: $3{\cdot}8 \times 6{\cdot}9$ would approximate to $4 \times 7 = 28$. Some symbols used for approximation are:

$>$, meaning 'greater than'
$<$, meaning 'less than'
$\geq$, meaning 'greater than or equal to'
$\leq$, meaning 'less than or equal to'
$\approx$ or $\doteqdot$, meaning 'approximately equal to'

APR Abbreviation for **annual percentage rate**.

apriori probabilities *See* **Bayes's theorem**.

apse Point on an orbit at which the motion of the orbiting object is at right angles to the central **radius vector** of the orbit. The apsidal distance (distance from an apse to the centre of the motion) equals the radius vector, at its maximum or minimum value.

arabic numeral Any of the characters 0 to 9, and so on. The term is used when distinguishing between values represented by those characters, and the same values represented by another character set. *Example*: the values represented by the characters 3, 4, 5, and 6 would be represented in **Roman numerals** by the characters III, IV, V and VI. *See also* **duodecimal**; **hexadecimal**; Appendix III (**Fibonacci**).

■**arc** Part of a curve cut at two specific points. The length of an arc of a circle, radius r, that subtends an angle of $\theta°$ at the centre is given by

$$\frac{2\pi r\theta}{360}$$

If θ is in radians, then the length of the arc $= r\theta$. [4/9/d] *See* **radian**; **subtend**.

arc- Prefix signifying an **inverse** function.

arc-cosine (acos) **Inverse** of the trigonometrical function **cosine**, signified by the symbol $\cos^{-1}$. *Example*: if $\cos 35° = 0{\cdot}81915$, then $\cos^{-1} 0{\cdot}81915 = 35°$. A **scientific calculator** will usually give $\cos^{-1}$.

architecture Relationship between the different parts of a **computer**.

arc-sine (asin) **Inverse** of the trigonometrical function **sine**, signified by the symbol $\sin^{-1}$. *Example*: if $\sin 0{\cdot}61086$ radians $= 0{\cdot}57357$, then $\sin^{-1} = 0{\cdot}61086$ radians. A **scientific calculator** will usually give $\sin^{-1}$.

arc-tangent (atan) **Inverse** of the trigonometrical function **tangent**, symbol $\tan^{-1}$. *See* **arc-cosine**; **arc-sine**.

are Unit of area in the **metric system**, equal to 100 m^2. *See also* **hectare**.

■**area** (A) Measure of the size of a surface, usually expressed as units squared. [4/5/f] The areas of some common figures are as follows, where l = length, h = height or **altitude**, r = radius:

square	l^2
rectangle	lh
parallelogram	lh
triangle	$\frac{1}{2}lh$
circle	πr^2
sphere	$4\pi r^2$
cone (curved surface)	πrs (s = slant height)
cone (total surface)	$\pi rs + \pi r^2$
cylinder (curved surface)	$2\pi rh$
cylinder (total surface)	$2\pi rh + 2\pi r^2$

A **scientific calculator** can be helpful in performing area calculations which include π, since such a calculator will have π as an inbuilt value. *Example*: calculate the total surface area of a cylinder with r = 6 m and h = 12 m. Key

$$2 \times \pi \times (6 \times 12) + 2 \times \pi \times (6\ x^2)$$

then =. The answer is 688·58 m^2. Note that it is important to include the × operators between 2 and π and between π and the bracketed operations; if this is not done, the answer will be wrong, or the calculator may supply an 'error' message.

arg Abbreviation for **argument**.

Argand diagram Set of **Cartesian coordinates** for representing **complex numbers**; the (vertical) y-axis gives the imaginary part of the number, and the (horizontal) x-axis gives the real part. *Example*: the complex number $x+iy$ is represented by the point plotted on the Argand diagram in Figure A9. The angle between the x-axis and a line from the origin to this point is called the **argument** of the complex number. The Argand diagram is named after the Swiss mathematician Jean Robert Argand (1768–1822).

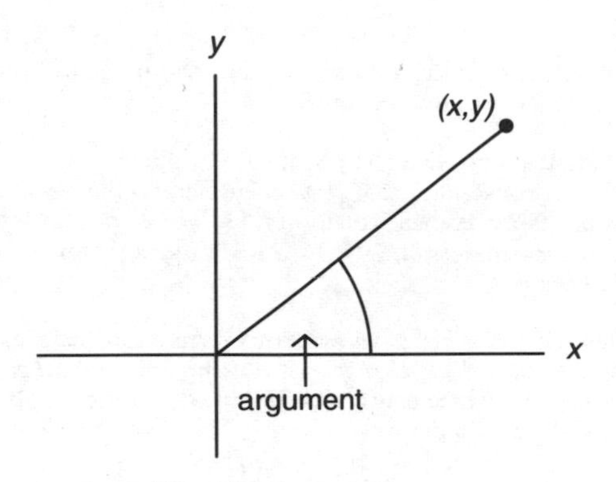

Fig. A9 *Argand diagram*

argument (arg) *1.* Variable term that forms part of a mathematical **function**; e.g. x in the function $y=5x^2$.
2. For the **complex number** $x+iy$, the angle whose tangent is $\frac{y}{x}$ (*see also* **Argand diagram**).

arithmetic Basic operations of adding, subtracting, multiplying and dividing, represented by the signs +, −, × and ÷ respectively.

■**arithmetic mean** Average of a collection of numbers obtained by dividing the sum of the numbers by the quantity of numbers. [5/4/f] [5/7/c] *Example*: the heights of five people in a group are 168·2; 185·6; 176·4; 169·3 and 182·3 cm respectively. Their mean height is:

$$\frac{168{\cdot}2+185{\cdot}6+176{\cdot}4+169{\cdot}3+182{\cdot}3}{5} = \frac{881{\cdot}8}{5} = 176{\cdot}36 \text{ cm}$$

Most **scientific calculators** with a statistical mode incorporate an arithmetic mean function.

■**arithmetic progression** (AP) Sequence of numbers for which there is a constant, d, such that the difference between any two successive terms is equal to d. [3/7/a] *Example*: for the arithmetic progression 2, 9, 16, 23, 30, the common difference is 7.

■**arithmetic series** Series whose terms form an **arithmetic progression**. [3/7/a] *Example*: if a is the first term, d is the common difference between the terms, and n the number of terms, the nth term is:

$$a+(n-1)d$$

and the sum is:

$$\tfrac{1}{2}n\,(2a+(n-1)d)$$

arithmetic unit Part of a computer's **central processor** that performs the arithmetical operations addition, subtraction, multiplication, and division.

ASCII Acronym for American Standard Code for Information Interchange, a widely-used computing code for representing characters and shapes in **binary**.

asin Abbreviation for **arc-sine**.

asset Financial value of a person or company's possessions. *See* **depreciation**.

associative Describing a two-stage mathematical operation of the type $a * (b * c)$ whose result does not depend on the order in which the operation is carried out. *Example*: the multiplication $3 \times (4 \times 5)$ gives the same result as $(3 \times 4) \times 5$, and thus multiplication is associative. Division, however, is not:

$$\frac{20}{(10/2)}$$

does not equal

$$\frac{(20/10)}{2}$$

It can be helpful to perform these operations, and look at the differences in results following from where the operators are placed, on a **scientific calculator**. The calculator operates using mathematical logic, and so quickly shows what difference is made by changes in the position of the operator. *See also* **commutative**.

astrolabe Instrument (now largely outdated) used to observe the positions and measure the altitudes of heavenly bodies.

astronomical unit (AU) Measure used for distances within the **Solar System**. One astronomical unit is equal to the mean distance between the Earth and the Sun $= 149{,}599{,}000$ km $= 92{,}953{,}274{\cdot}51$ **miles**.

asymmetrical Describing any configuration which does not possess the quality of **symmetry**. *Example*: a pair of scales with a much heavier weight on one side than the other is asymmetrical.

asymptote In geometry and coordinate geometry, a line that a curve approaches more and more closely but never reaches. *Example*: Figure A10 shows the graph for the equation $y=\frac{1}{(x-1)}$, for $x \geq 1$. Line $x=1$ and the x-axis are both asymptotic.

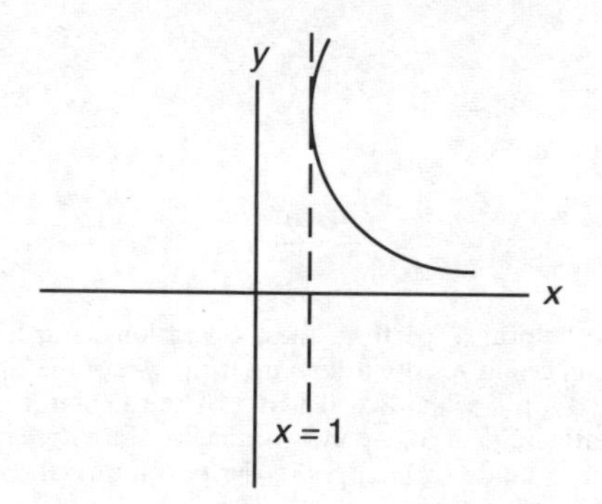

Fig. A10 *asymptote*

atan Abbreviation for **arc-tangent**.

ATM Abbreviation for **automatic teller machine**.

atto- (a) Prefix in the **metric system** signifying $\times 10^{-18}$.

AU Abbreviation of **astronomical unit**.

automatic teller machine (ATM) Computerised machine, made available by banks and building societies to their account holders, that allows the account holders to withdraw cash automatically using **cash cards** from the account the cash card is issued for.

■**average** Number that is representative of a collection of numbers; e.g. an **arithmetic mean, mode** or **median**. [5/4/f]

axes Plural of **axis**.

axiom *1.* An established principle or self-evident truth. *See* **Euclidean geometry**. *2.* An assumption on which a logical argument is founded.

■**axis** In mathematics, a line of significant reference for a graph or figure; e.g. x- and y-axes in **Cartesian coordinates**. [5/4/d]

axis of symmetry Line joining the centres of the ends of a three-dimensional object. *Example*: the axis of symmetry in a cylinder is the line that can be drawn through the centres of both the circular ends of the cylinder.

azimuth In astronomy, mathematics and surveying, the angle between the vertical plane containing the line of sight to an observed object, and the plane of the **meridian**.

B

B Number 11 in the **duodecimal** number system.

B Number 11 in the **hexadecimal** number system.

backing store Computer store that is larger than the main (immediate access) memory, but with a longer access time.

ballistics Study of projectiles moving under the force of gravity only.

■**bar chart** Graph that has vertical or horizontal bars whose lengths are proportional to the quantities they represent. [5/3/c] *See also* **graphical calculator**; **pie chart**.

■**bar line graph** Graph where the data is represented either by a line or a narrow bar drawn to scale. The length of each line is proportional to the number value of the data represented. [5/4/c]

barrel Unit of volume. In the oil industry, 1 barrel = about 159 litres (35 gallons); in brewing, 1 barrel = 32 gallons.

base *1.* In geometry, the horizontal line upon which a geometric figure stands. *2.* In mathematics, the starting number for a numerical or logarithmic system; e.g. binary numerals have the base 2, common **logarithms** are to the base 10. *3.* In mathematics, number on which an **exponent** operates; e.g. in 52, 5 is the base (and 2 the **exponent**).

BASIC Acronym for Beginner's All-Purpose Symbolic

Instruction Code, a widely-used computer programming language, most frequently on **PC**-based systems.

Bayes's theorem Equation used in **probability** which allows the user to calculate the probability of an event occurring, provided that another event whose probability is already known has occurred. The probability is $P(a_i)$ of the event a_i occurring when a_i is one of n mutually exclusive events in the series

$$a_1 \; a_2 \; a_3 \; .... a_n$$

Then if E is an observable event $P\left(\frac{E}{a_i}\right)$ is the probability that E will occur assuming that a_i has occurred and $P\left(\frac{a_i}{E}\right)$ is the probability of a_i occurring if E has occurred. Then $P(a_i)$ are called **apriori probabilities** and $P\left(\frac{a_i}{E}\right)$ the **aposteriori probabilities**. Generally $P(a_i)$ are not known. The equation fully stated is as follows:

$$P\left(\frac{a_i}{E}\right) = \frac{P(a_i) \,.\, P\left(\frac{E}{a_i}\right)}{\sum_{j=1}^{n} P(a_j) \,.\, P(E/a_j)}$$

■ **bearing** In surveying and telecommunications, the horizontal angle between a line and a reference direction, often measured clockwise from North. [4/3/c] [4/6/h] *Example*: in Figure B1, the bearing is 060°, or N 60° E.

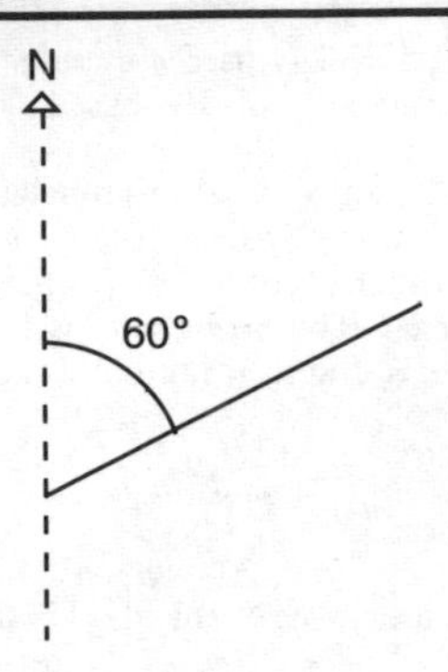

Fig. B1 *bearing*

Bernoulli's theorem Theorem which states that at any point in a tube through which a liquid is flowing the sum of the potential, kinetic and pressure energies is constant. Alternative name: Bernoulli's principle. It was named after the Swiss mathematician and physicist Daniel Bernoulli (1700 – 1782).

■ **bilateral symmetry** Type of symmetry in which a shape is symmetrical about a single plane (each half is a mirror image of the other). [4/3/b] *Example*: creatures with spines, such as human beings, are bilaterally symmetrical. See also **radial symmetry**; **rotational symmetry**.

billion Number now generally accepted as being equivalent to 1,000 million (10^9). Formerly in Britain a billion was regarded as a million million (10^{12}).

bimodal Having two **modes**. The term often occurs in descriptions of statistical distributions when two items in a set occur with the same, or almost the same, frequency. *Example*: in the set {1,1,1,2,2,3,3,3,4,5,6} both 1 and 3 are modes of the set.

BIN Abbreviation for **binary notation**.

■ **binary code** Alternative name for **binary notation**. [2/4/c,f]

binary digit The numbers 1 and 0 used in **binary notation**. Often abbreviated to **bit**.

■ **binary notation** (BIN) Number system to the base 2, involving only two digits, 0 and 1. Instead of the units, tens, hundreds etc. of the decimal system, units, twos, fours etc. are used. Thus, for example,

decimal		*binary*
1	=	1
2	=	10
3	=	11
4	=	100
5	=	101
8	=	1000
9	=	1001
20	=	10100
30	=	11110

Binary notation is important in electronics and computers because the 0 and 1 can be represented by a circuit being 'on' or 'off'. Most **scientific calculators** have a binary mode which allows calculations to be made in this number base, and conversions to be made from binary to other number bases. [2/4/c,f] Alternative name: **binary code, binary system**.

binary operation Combining of two terms to produce a third; all four of the arithmetic operations +, −, × and ÷ are binary since they combine two terms to produce another. *Example*: $3+4=7$, $4-3=1$, $3\times4=12$, $3\div4=0{\cdot}75$.

■ **binary system** Alternative name for **binary notation**. [2/4/c,f]

binomial In mathematics, a **polynomial** which has two variables, for example $3a\times4b$, $x-y$, $(2x+3y)^2$.

■ **binomial coefficients Coefficients** or numbers of the terms which arise in any **binomial expansion**. [3/6/c] *Example*: in the binomial expression $(1+x)^n$ the coefficients of the powers of x are binomial coefficients.

■ **binomial distribution** Statistical distribution generated by a binomial expansion and given the probability of an event occurring. [3/6/c] [3/7/e] *Example*: if the probability of an event happening in a single attempt is p, the probability q it will not happen equals $1-p$; if there are n attempts the probability of an event happening b times is:

$$^{n}C_{b}\,.\,p^{b}\,.\,q^{n-b}$$

The distribution would be:

$$p^{n}+{}^{n}C_{1}\,.\,p^{n-1}q+{}^{n}C_{2}\,.\,p^{n-2}q^{2}+\ldots{}^{n}C_{b}\,.\,p^{b}q^{n-b}$$

binomial expansion Result of multiplying out a **binomial** expression supplied to a given **power**, or of multiplying out n binomial expressions. *Example*: the expansion of the binomial $(1+x)^3$ is

$$(1+x)(1+x)(1+x)$$

which gives

$$(1+x+x+x^2)(1+x)$$

which gives

$$(1+2x+x^2)(1+x)$$

which gives

$$1+2x+x^2+x+2x^2+x^3$$

which gives

$$1+3x+3x^2+x^3$$

The **binomial coefficients** (the coefficients of x) in this expansion are 3, 3, 1. As can be seen by considering **Pascal's triangle**, these coefficients correspond to the grouping on line 4 from the top of the triangle. Use of the triangle makes it possible to expand a binomial to any power without using the lengthy multiplying out employed in the example; the coefficients can be found on the relevant line of the triangle.

binomials, multiplication of *See* **product**.

bisect To cut in half. A line or an angle is cut into equal parts when it is bisected, as in Figure B2, where an angle of 60° is bisected.

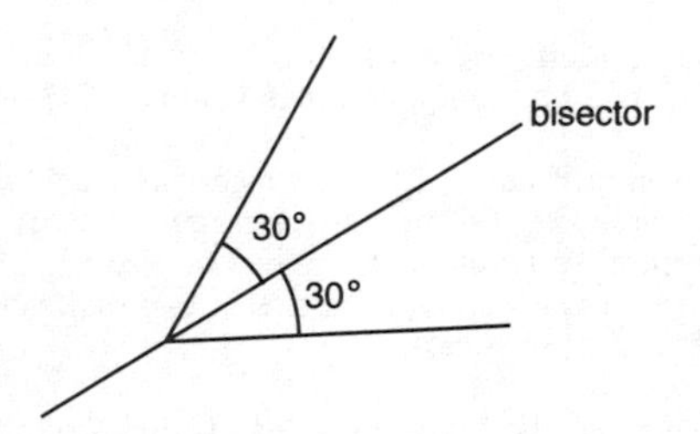

Fig. B2 *bisect*

bisector Line that bisects another.

bit Amount of information that is required to express choice between two possibilities. The term is commonly applied to a single digit of **binary notation** in a computer. The word is an abbreviation of **binary digit**.

■ **block graph** Graph of a set of frequencies denoted by a set of rectangles whose areas are proportional to the frequencies. If the width of each block is the same then the outcome is a column graph. [5/3/c]

Bode's law Probably coincidental numerical sequence that gives the distances of the planets of the Solar System from the Sun. Adding 4 to each member of the series 0, 3, 6, 12, 24, 48, . . . 384 (in which each term is twice the previous one) gives the new series 4, 7, 10, 16, 28, 52, . . . 388. If the Earth is identified with the number 10, then

4 = Mercury
7 = Venus
16 = Mars
52 = Jupiter

Neptune's position does not fit this 'law,' which was named after the German astronomer Johann Bode (1747–1826).

Boolean algebra Branch of logic, which uses algebraic symbols, employed in computer programs. Its chief operations are AND and OR, applied to elements that are *true* or *false*. It was named after the British mathematician George Boole (1815–1864).

borrower Person who takes out a **loan** from a financial body, such as a bank or building society.

boundary Limit line, usually placed around a certain area.

box and whisker plot Diagram showing some important information about the distribution of statistical data. A box is drawn between the upper and lower quartiles and this includes the median. Two lines (or whiskers) are drawn from the box representing the extreme values at each side of the quartiles, as in Figure B3. Alternative term: **boxplot**.

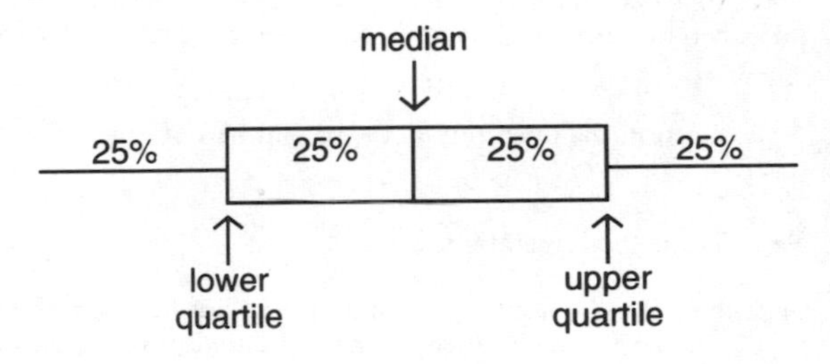

Fig. B3 *box and whisker plot*

bracket Sign used in mathematics to enclose a value or set of values. The three most commonly used kinds are:

1. curved bracket, as in $1 \times (7-8)$;
2. fretted bracket, used to signify sets, as in $\{1, 2, 3, 4\}$;
3. squared bracket, used to signify intervals, as in $[a, b]$.

Values enclosed by round brackets are treated as a single term in any operation. *Example*:

$$4 \times (-10+8) = 4 \times (-2) = -8$$

Note: ignoring the brackets gives an incorrect answer:

$$4 \times (-10) = -40$$

$$-40 + 8 = -32$$

Pairs of rounded brackets can also be used, as in

$$8 + ((6 - (10 + 3))$$

Here, the brackets also determine the order in which the operations demanded by the + and − signs take place. The add operation in $(10+3)$ should be carried out first, to give

$$8 + (6 - 13)$$

Now, perform the operation in the second pair of brackets to give

$$8 + (-7)$$

Perform the final operation: $8 - 7 = 1$

Separate expressions enclosed within round brackets can also be placed next to each other to signify multiplication of the expressions, as in

$$(2a + b)(4a - 3) = 8a^2 - 6a + 4ab - 3b$$

or, substituting 3 for a and 5 for b

$$\begin{aligned} ((2 \times 3) + 5))((4 \times 3) - 3)) &= ((6) + 5)((12) - 3) \\ &= 72 - 18 + 60 - 15 \\ &= 99 \end{aligned}$$

Most **scientific calculators** have round bracket keys which allow the entry of expressions which include brackets, and follow the order of calculation for expressions within brackets described above. When using such calculators, however, it is important to place an operator before the opening bracket, otherwise the calculation will not result in a correct answer. *Example*: if a solution is sought to 12(18 − 10), the expression should be keyed as 12 × (18 − 10).

British unit System of measurement based on the **yard**, which is in turn based on the **f.p.s.** system.

bug Error in a computer **program** or a fault in **hardware**.

bureau de change Specialized **currency** exchange office. Bureaux de change (the plural) can usually be found in banks, hotels, airports and railway stations and as small shops in large cities.

bushel Unit for measuring dry goods by volume, equal to 4 pecks and equivalent to 8 gallons or 36·369 litres in Britain, or 35·238 litres in the USA.

byte Single unit of information in a **computer**, usually a group of 8 **bits**.

C

C *1.* Number 12 in the **hexadecimal** number system. *2.* Number 100 in the **Roman numeral** system. *3.* Symbol for **Celsius**.

c Symbol for **centi-**, $\times 10^{-2}$.

calculator Manual, mechanical or electronic device for performing arithmetical and mathematical calculations. Calculators of one kind or another have been in existence for thousands of years. Manual devices, such as the **abacus**, are still in use, while others, such as the **slide rule**, have now fallen into disuse. There have been extremely complex mechanical devices, such as Babbage's **analytical engine**, but handheld **digital** electronic calculators are now probably the most widely-used type of calculator. These can range from the extremely simple, which will add, subtract, divide, multiply and give percentages, through the more complex (generally described as **scientific calculators**) which can, for example, calculate **common logarithms**, **Napierian logarithms** and **inverse functions**, to **graphical calculators** which will, for example, plot **sine** curves. The **computer** is essentially an electronic calculator. *See* Appendix III, **Babbage**; **Leibniz**; **Pascal**.

calculus Branch of mathematics that deals with integration and **differentiation**, from which the terms **integral calculus** and **differential calculus** are derived. The basic approaches to integration and differentiation underlying what we now understand as calculus were developed in the 17th century principally by the English scientist Isaac Newton (see

Newtonian mechanics; **Newtonian method**), the German mathematician and philosopher Gottfried Liebniz (1646–1716) and by the French mathematician Pierre de Fermat (1601–65). All three worked independently of one another, and their respective contributions to calculus are, therefore, distinctive. All three were, however, attempting to find mathematical ways to deal with infinitesimals. These are potentially infinite subdivisions of changes in the state of a mathematical entity (e.g. the progressive motion of a curve) where the difference between one state (e.g. a curve at a single point) and a second state (the same curve at a second point) are treated as a succession of infinitely small increases or decreases. Infinitesimals had, until the advent of calculus, been one of the principal methods employed by mathematicians to analyse phenomena such as **accleleration**: how is the path of an accelerating (or decelerating) body analysed; what is the speed of a body at point x_1 in its motion compared with its speed at point x_2 in its motion? The development of calculus created a consistent mathematical language for describing such changes, and also contributed significantly to associated questions, such as the method of finding the **tangent** to any curve at any point, the method of measuring the length of a curve, and the method of finding the **area** under a curve. *See also* **derivative**; **function**; **mid-ordinate rule**; **Simpson's rule**; **trapezoidal rule**; Appendix III.

calendar year Time it takes the Earth to orbit the Sun, averaging 365·25 mean **solar days** and organized as three years of 365 days followed by a **leap year** of 366 days.

calibration Method of putting a scale on a scientific instrument, usually by checking it against fixed quantities or standards.

cancellation Method of simplifying a fraction by dividing the top and bottom (the numerator and denominator) by a common factor. For example, dividing both the numerator and denominator of 18/24 by 6 produces the simpler fraction 3/4.

candela (cd) **SI unit** of measure of luminous intensity.

cardinal number *1.* Number of elements in a set. For example, all sets with 5 elements have the cardinal number 5. *2.* An ordinary counting number. *See also* **ordinal number**.

cardioid Heart-like shape traced by a point on the circumference of a circle that rolls round another circle of the same radius. The polar equation is $r=2a\ (1-\cos\theta)$ where r is the radius vector and θ the vectorial angle. There are always three tangents with the same gradient on this curve.

card punch Machine for punching coded sets of holes in punched cards, to be fed through a **card reader** for inputting **data** into a computer.

card reader Computer **input device** that reads **data** off punched cards.

■ **Cartesian coordinates** System for locating a point, P, by specifying its distance from axes at right-angles, which intersect at a point, O, called the origin. For a point on a plane, the distance from the horizontal or x-axis is called the **ordinate** of P; the distance from the y-axis is called the **abscissa**. The point's Cartesian coordinates are (x, y). The system is named after the French philosopher and mathematician René Descartes (1596–1650), and is the one most frequently used for graphically locating a point in **coordinate geometry**. *See also* **polar coordinates**. [3/6/c]

cash card Plastic card issued usually by a bank or building society enabling a user to withdraw cash from the account for which the card is issued by means of an **automatic teller machine**. Alternative term: **cheque guarantee card**.

catenary In mathematics, curve formed by a chain or heavy cable supported at its ends. It is expressed by the equation

$$y=\tfrac{1}{2}a(e^{x/a}+e^{-x/a})=a\cosh(x/a)$$

It is therefore the graph of $y=\cosh x$, which is in fact a **hyperbolic function**.

cc Abbreviation for **cubic centimetre**.

cd Abbreviation for **candela**.

celestial equator Circle in which the plane of the Earth's **equator** meets the **celestial sphere**. The celestial North Pole is near the North Star and the celestial South Pole is near the Southern Cross star constellation.

celestial sphere Imaginary sphere to whose inner surface the heavenly bodies appear to be attached. The celestial poles are immediately above the Earth's poles, and the sphere is bisected by the **celestial equator**.

Celsius scale Temperature scale on which the freezing point of water is 0 °C and the boiling point is 100 °C. It is the same as the formerly used **centigrade scale**, and one degree Celsius is equal to one unit on the **kelvin** scale. To convert a Celsius temperature to the **Fahrenheit** scale, multiply by 9/5 and add 32. *Example*: 25 °C = 25 × 9/5 = 225/5 = 45; 45 + 32 = 77 °F.

centenarian Person reaching the age of 100.

centesimal Describing the hundredth part of a quantity or value.

centi- Prefix signifying $\times 10^{-2}$, as in **centimetre**.

centigrade scale Former name of the **Celsius scale**.

■ **centimetre** One hundredth part of a metre, abbreviation cm. [2/5/i]

central processor Part of a **digital computer** which controls and coordinates all the other activities of the machine, and performs logical processes on **data** loaded into it according to **program** instructions it holds.

centre Point in a **circle** equidistant from every point on the **circumference**.

centre of curvature Geometric centre of a spherical mirror.

■ **centre of enlargement** Point from which an enlargement takes place. The centre of enlargement need not be at the origin. [4/7/e] *Example*: consider the triangle XYZ in Figure C1: the centre of enlargement C is found by joining XX′ and ZZ′ with straight lines and extrapolating them until they meet. C can be found by solving a matrix equation

$$\begin{pmatrix} \mathrm{E} & \mathrm{O} \\ \mathrm{O} & \mathrm{E} \end{pmatrix}\begin{pmatrix} x \\ y \end{pmatrix} + \begin{pmatrix} a \\ b \end{pmatrix} = \begin{pmatrix} x \\ y \end{pmatrix}$$

where E is the enlargement factor, x and y are the coordinates of C and $\begin{pmatrix} a \\ b \end{pmatrix}$ is a translation vector.

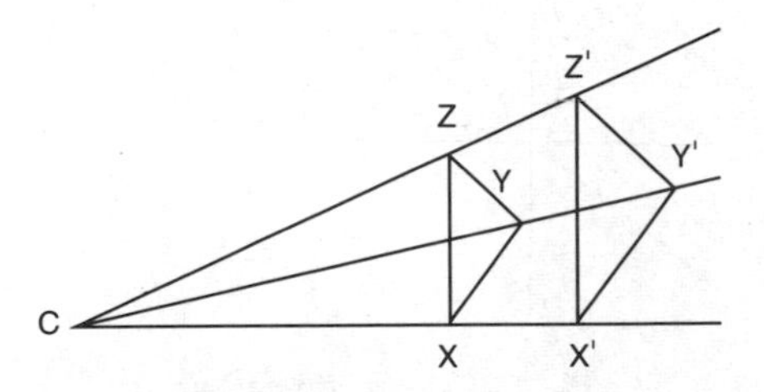

Fig. C1 *centre of enlargement*

centre of mass Point at which the whole **mass** of an object may be considered to be concentrated. Alternative name: centre of gravity.

■**centre of rotation** Point about which a rotation or turning occurs. [4/4/e] *Example*: in Figure C2, point A rotates to position B in an **anticlockwise** direction:

centroid Point within an irregular shape or object at which the **centre of mass** would be if the shape or object were of uniform density. It is coincident with the centre of mass for a symmetrical figure.

cevian Line joining the **vertex** of a triangle to a point on the side of the triangle opposite the vertex. For example, the **altitude** of a triangle is a cevian.

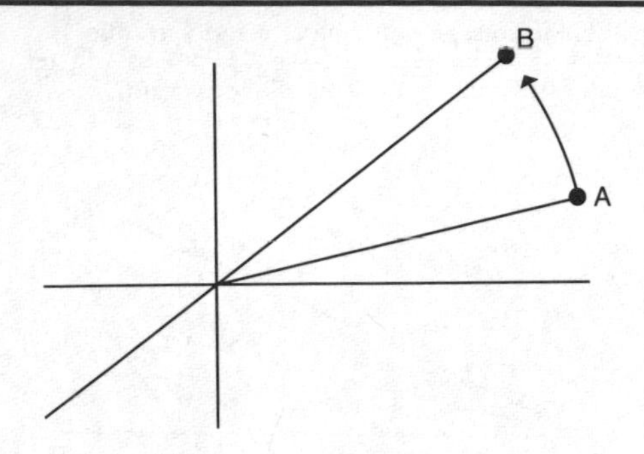

Fig. C2 *centre of rotation*

c.g.s. system System of measurement based on the **centimetre**, **gram** and **second**. **SI units** have largely replaced the c.g.s. system. See also **f.p.s. system**.

■ **changing the subject of a formula** The subject of a formula is the single term on one side of the equation. *Example*: in the formula $x = lb$ for the area of a rectangle (where x is area, l is length and b is breadth), the subject is x. The subject can be changed to l by dividing both sides of the formula by b, to give $x/b = l$. A more complicated example is the formula for the area of a sphere: $x = 4\pi r^2$. Changing the subject to r gives $r = \frac{1}{2}\sqrt{x/\pi}$. [3/8/a] *Example*: Transpose the formula

$$D^3 = \frac{4a^2b}{C}$$

to make C the subject: (1.) Multiply both sides by C:

$$CD^3 = \frac{\not{C}4a^2b}{\not{C}}$$

(2.) Simplify and divide both sides by D^3:

$$\frac{C\not{D}^3}{\not{D}^3} = \frac{4a^2b}{D^3}$$

(3.) Therefore:

$$C = \frac{4a^2b}{D^3}$$

characteristic Integer part of a **logarithm**. *Example*: the logarithm (to the base 10) of 200 is 2·3010, in which 2 is the characteristic and ·3010 is the **mantissa**.

character recognition Method of inputting information (e.g. to a computer) by optically or magnetically scanning printed or hand-written characters. *See also* **bar code**.

cheque guarantee card Plastic card issued by banks and building societies which guarantees that a cheque issued by a card holder will be met in full. Such cards currently carry spending limits of £50 and £100; most also function as **cash cards**.

■ **chord** Straight line that joins two points on a curve. [4/9/d] *Example*: in Figure C3, the chord BC joins the two points B and C on the circumference of a circle.

■ **circle** Perfectly round plane figure contained by a line, the **circumference**, which is equidistant from a fixed point within it, the **centre**. In **coordinate geometry**, the equation for

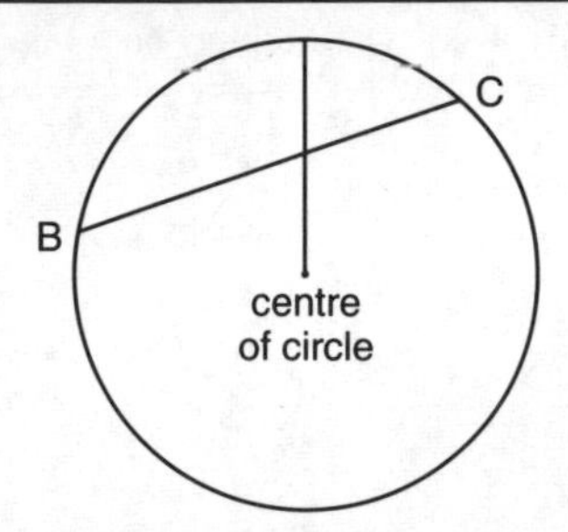

Fig. C3 *chord*

finding the radius of a circle is:

$$(x-a)^2+(y-b)^2=r^2$$

where r is the radius and the coordinates of the centre are (a,b). [4/2/a]

■ **circumference** Boundary of a circle, equal in length to πd or $2\pi r$, where d = the diameter and r the radius of the circle. The area of a circle is calculated using the formula πr^2. *Example*: use a calculator to find the area of a circle with circumference 7·3 m. Key π, then ×, then 7·3, then x^2, then =. The answer is 167·41547 m². [4/2/a]

circumscribed Drawn around, as a **circumscribed circle** is drawn around a **polygon**.

circumscribed circle Circle that passes through the vertices of a polygon. The centre of the circumscribed circle is at the point where the perpendicular bisectors of the sides of the polygon intersect, as in Figure C4, where the perpendicular bisectors of the sides of triangle ABC intersect at the centre of the circle circumscribing triangle ABC.

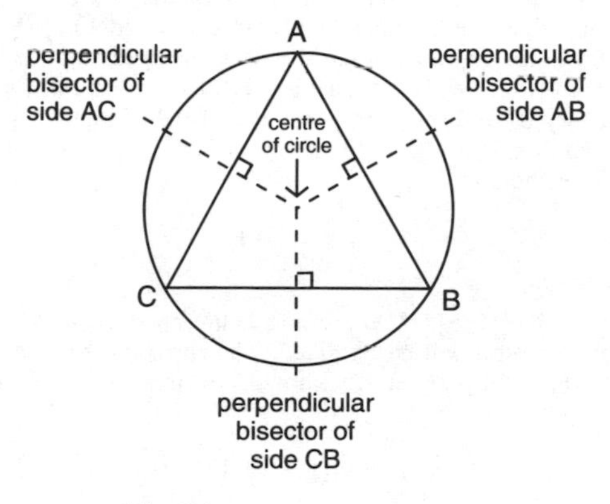

Fig. C4 *circumscribed circle*

The radius r is given by the formula:

$$r = \frac{abc}{4x}$$

where a, b, c are the lengths of the sides and x is the area of the triangle. The area of the triangle can be found by using the formula

$$\sqrt{s(s-a)(s-b)(s-c)}$$

where s is the semi-perimeter, i.e.

$$\frac{(a+b+c)}{2}$$

Example: using a calculator, find the radius of a circle circumscribed on a triangle with sides 6, 8 and 10 m. From the formulae supplied, it can be seen that it is necessary to know the area of the triangle to find the radius, and that it is necessary to find the semi-perimeter of the triangle to calculate its area. The first step, therefore, is to calculate the semi-perimeter, =

$$\frac{(6+8+10)}{2}$$

Key $(6+8+10) \div 2$. If the brackets are missed out, an incorrect result will be obtained. The correct answer is 12. Since the semi-perimeter is known, the area can be found with

$$\sqrt{12(12-6)(12-8)(12-10)}$$

Key $\sqrt{}(12 \times (12-6) \times (12-8) \times (12-10))$

The correct answer is 24 m^2. Now that the area is known, the radius of the circumscribed circle can be found with

$$\frac{6 \times 8 \times 10}{4 \times 24}$$

Key $(6 \times 8 \times 10) \div (4 \times 24)$. The correct answer is 5 m.

■**class interval** Limit of a numerical group or set of data. [5/8/c] *Example*: 0–99, 100–199, 200–299 are all class intervals.

clockwise Describing **rotation** in the direction normally followed by the hands of an **analog** clock or watch. *Compare* **anticlockwise**.

closed curve Continuous curve with no ends; a loop. *Example*: for the curve $x=a \,.\, \sin\theta$ and $y=b \,.\, \cos\theta$ and $0 \leq \theta \leq 2\pi$ both end points are $(a,0)$ and the curve becomes an ellipse.

closed interval Set of all numbers that lie between (and include) two other numbers, given by $a \leq x \leq b$ where x represents the numbers. For example all the numbers between 3 and 9 are the closed interval given by $\{x\colon 3 \leq x \leq 9\}$.

closed set Set in which the combination of any two members of a set by an operation gives a member of the original set. *Example*: under the operation of addition the set of natural numbers $\{1, 2, 3, 4, 5, \ldots\}$ is closed because adding any of the numbers together results in another number that is already in the set.

COBOL Acronym of COmmon Business Oriented Language, a computer programming language designed for commercial use.

coefficient In an algebraic expression, the numerical factor by which the variable is multiplied. *Example*: in $5xy$, the coefficient of xy is 5.

coefficient of restitution Ratio of the velocity of an object after impact to that before impact, usually denoted by the symbol e. *Example*: if an object is dropped from a height h_1 above an elastic plane, and rebounds to a height h_2, then the coefficient of restitution is equal to $\sqrt{h_2/h_1}$. See ***e***.

collinear *1.* Describing points that lie on the same straight line. *2.* Describing planes which share the same axis i.e. coaxial planes.

column Elements arranged vertically, such as those in a **determinant** or **matrix**.

combination In mathematics, any selection of a given number of objects from a set, irrespective of their order. The number of combinations of r objects that can be obtained from a set of n objects (usually written ^nC_r) is:

$$\frac{n!}{r!(n-r)!}$$

Pascal's triangle is a table of combinations.

■ **combined probability** If a objects can be taken from n objects the possible number of selections is written as ^nC_a. This is known as the combined probability. The value of $^nC_a=$

$$\frac{n!}{a!(n-a)!}.$$

The ! is the **factorial** operator. [5/6/f] *Example*: if 3 balls can be selected from 5 balls, then the possible number of selections $= {}^5C_3 =$

$$\frac{5!}{3!(5-3)!}=10$$

On a calculator with an $x!$ key and brackets (. . . . . .) keys, key

$$5x! \div (3! \times ((5-3)x!)) =$$

On a calculator with a ^nC_r key 5 ^nC_r3 = .

■**common denominator** Denominator which is a common multiple of the denominators of fractions with different values, allowing ease of arithmetic operations using fractions. [2/4/k] *Example*: to subtract 1/3 from 1/2 the fractions are first assigned the common denominator of 6, becoming 2/6 and 3/6, and subtracted to give 1/6. Using the same method they can be added to give 5/6.

common difference Difference between any two consecutive terms of an **arithmetic progression**.

common logarithm (log) **Logarithm** to the base 10. Most **scientific calculators** will return values for common logarithms. Alternatively, they can be found listed on logarithm tables. *Compare* **Napierian logarithm**.

commutative Describing a mathematical operation of the type $a * b$ whose result does not depend on the order in which the operation is carried out. *Example*: the addition $5+4$ has the same result as $4+5$, so addition is commutative. Subtraction is not; $5-4$ is not the same as $4-5$. Multiplication, translation and rotation in a plane are also commutative. *See also* **associative**.

compass Apparatus for finding direction (parallel to the Earth's surface), usually by alignment with the Earth's magnetic field.

compasses Instrument consisting of two hinged arms, one of which ends in a sharp point and the other in a drawing point. It is very useful in geometric **construction**, principally for drawing **circles** of a given radius, and for drawing **arcs**.

compatibility (of matrices) Matrices can only be added or subtracted if they are the same order. Two matrices can only be multiplied together if the number of columns in one is equal to the number of rows in the other. Therefore the matrices have to be compatible or consistent with each other.

compiler Computer **program** that converts a source language into **machine code** (readable by the computer).

complement *1.* In set theory, the set of all members of the universal set that are not members of *S*, usually written as *S*′. *2.* In computing, the number obtained by subtracting each digit from one less than the number base and adding one to the result. *Example*: if the computer were using the decimal system (base 10), the complement results from subtracting each digit from 9 (1 less than base 10) and adding 1 to the answer; thus the complement of 123 is $(999-123)+1=876+1=877$. *3.* In geometry, an angle which, when added to another, totals 90°.

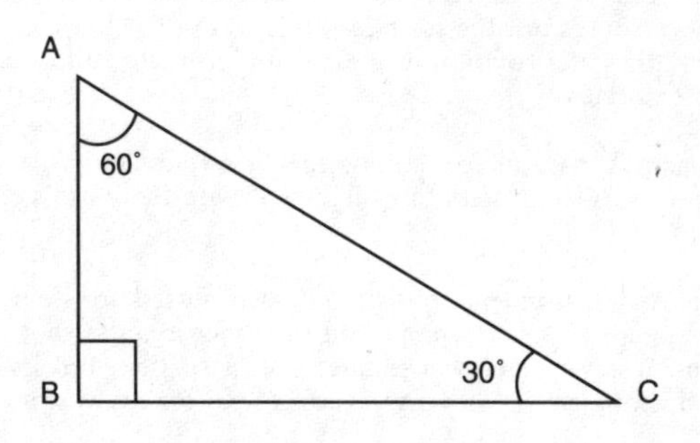

Fig. C5 *complementary angles*

■**complementary angle** One of a pair of angles that add to 90°, for example the angles other than the right angle in a right-angled triangle. Each of the two angles is the **complement** of the other. [4/5/b] *Example*: see Figure C5, where angles BAC and BCA in the right-angled triangle ABC together equal 90°, and are therefore complementary angles.

completing the square Method of solving a **quadratic equation** by getting it into the form $(x+a)^2=b$, which is easy to solve because $x+a=\sqrt{b}$, or $x=\sqrt{b}-a$. *Example*:

$$2x^2+8x-4=0$$
$$2x^2+8x=4$$
$$x^2+4x=2$$
$$(x+2)^2-4=2$$
$$(x+2)^2=6$$
$$x+2=\sqrt{6}$$
$$\begin{aligned} x&=\sqrt{6}-2\\ &=2{\cdot}45-2\\ &=0{\cdot}45 \text{ or } -4{\cdot}45 \end{aligned}$$

complex number Number written in the form $x+iy$, where x and y are **real numbers** and i is the square root of -1 (i.e. $i^2=-1$), an **imaginary number**. Some **scientific calculators** will perform calculations using complex numbers. *See also* **Argand diagram**.

■**component** In mathematics, the resolved part of a **vector** quantity in a particular direction, usually one of a pair at right angles. *Example*: in Figure C6, where vector **v** is shown with angle θ, the vertical component $= \mathbf{v} \sin \theta$, and the horizontal component $= \mathbf{v} \cos \theta$. [4/9/a].

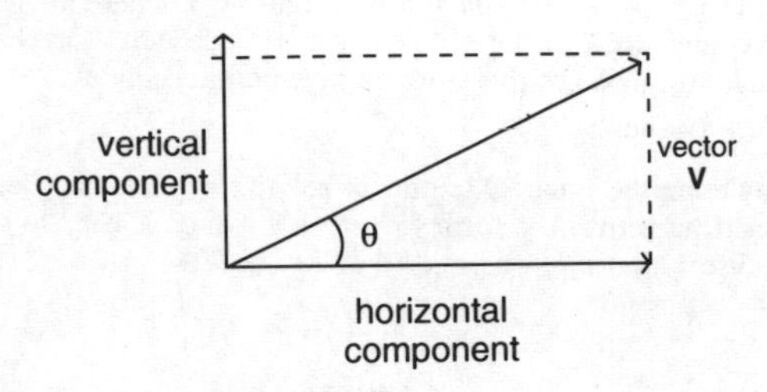

Fig. C6 *component*

composite function If f is a function of x $[f(x)]$ and g is a function of x $[g(x)]$ then the composite function $gf(x)$ can be found. Composite functions are very rarely commutative so $fg(x)$ does not equal $gf(x)$.

compound interest Interest calculated in such a way as to take into account interest previously earned. After n years a sum S invested at x per cent compound interest is worth $S(1+x/100)^n$. *Example*: use a pocket calculator to find out what the compound interest would be on £100 invested at 8·6% interest over 5 years. Following the formula, write the problem down with the values supplied:

$$100(1+8{\cdot}6/100)^5$$

Key $8{\cdot}6 \div 100$; press $=$. Rewrite the problem as

$$100(1+0{\cdot}086)^5$$

Now rewrite it as

$$100(1{\cdot}086)^5$$

Key 1·086; press the x^y key; key 5; press =. This gives 1·5105. With this number still on the calculator screen, key × 100; then press =, which gives the answer: £151·05.

computer Electronic device that can accept **data**, apply a series of logical operations to it (obeying a **program**), and supply the results of these operations via a **peripheral** such as a **printer**. Computers are essentially electronic **calculators** which use **binary notation** to perform their operations. Many different kinds of computer are now in use. They range from handheld devices, through **laptops** and desktop machines (**PC**s) to much larger machines, known until recently by the broad term **mainframe**. The main differences between them are the complexity of their **architecture**, the amount of memory they have available, and their speed of operation. Very fast **super computers** exist, such as the Cray 3, with clock speeds of around 2 GHz (PC-based system speeds are measured in MHz, $\times 10^{-3}$ slower). Super computers are generally used for highly complex operations, such as the storage and manipulation of weather data. *See* **analog computer**; **digital computer**.

concave Describing a curve or surface that is hollowed inwards (like the inside of a bowl or saucer). If any interior angle in a **polygon** is greater than 180° then it is concave.

concentric Describing a number of circles with a common centre, or cylinders with a common axis.

■**concurrent** Describing lines which pass through the same point. For example, parallel lines are never concurrent because they are always the same distance apart. [4/5/b]

■**cone** Solid figure, with a circular flat base, that comes to a point. Its volume equals $1/3\pi r^2 h$, where r is the radius of the base and h is the perpendicular height. [4/2/a] The area of the curved surface of a right cone is πrs, where s is the slant height; the total area is $\pi rs + \pi r^2$.

confidence interval Interval between which it is assumed that data is reasonably accurate in a statistical sample. It may be part of a certain sample of data.

■**congruent** *1.* In mathematics, describing plane or solid figures that have the same size and shape. Congruent figures can be made to overlap (after **rotation** or **reflection**, if necessary) so that they 'fit' each other perfectly. [4/4/b] *2.* In **modular arithmetic**, x and y are congruent in modulo n if the difference $x - y$ can be exactly divided by n.

conic section Curve produced when a plane cuts through a **cone**. Depending on the angle of cut, the section is a circle, ellipse, parabola or hyperbola. Each of these curves is the **locus** of a point that moves in such a way that the ratio of its distance from a fixed point (the **focus**) to its distance from a fixed line (the **directrix**) is a constant (called the **eccentricity**).

conjugate *1.* In geometry, describing a pair of angles whose sum is 360°. *2.* Of the **complex number** $x + iy$, the number $x - iy$ (represented by a reflection on the opposite side of the real axis on an **Argand diagram**). The product $x^2 + y^2$ is a real number.

constant Quantity that remains the same in all circumstances. *Example*: (a) the acceleration of free fall is a physical constant; (b) in the expression $4y = 5x^2$, the numbers 4 and 5 are constants and x and y are **variables**.

■**construction** *1.* In geometry, a drawing made by following certain rules. [4/7/b] *Example*: Figure C7 shows how a perpendicular line can be constructed from a given point on a horizontal line using a pair of **compasses** and a ruler:

(1). select any point on a horizontal line and put the sharp point of the compasses at that point;

(2). draw an arc with the compasses using the point you selected so that the arc cuts the horizontal line and can be seen directly above and directly below the line;

(3). keeping your compass points the same distance apart, place the sharp point where the arc cuts the horizontal, and then draw an arc from that point so that it cuts your previous arc directly above and below the line;

(4). using a ruler, join the points where the two arcs have cut one another. The resulting line should now be at a precise right angle to your original horizontal.

2. In mathematics, a way of creating simple equations from given rules. The equations can then be solved to find the unknown quantity. *Example*: what is the number which, if you subtract 6 from it and then multiply it by 2 gives an answer of 8? Let the unknown number be x. Then:

$$2(x-6)=8$$

$$x-6=4$$

$$x=4+6$$

$$x=10$$

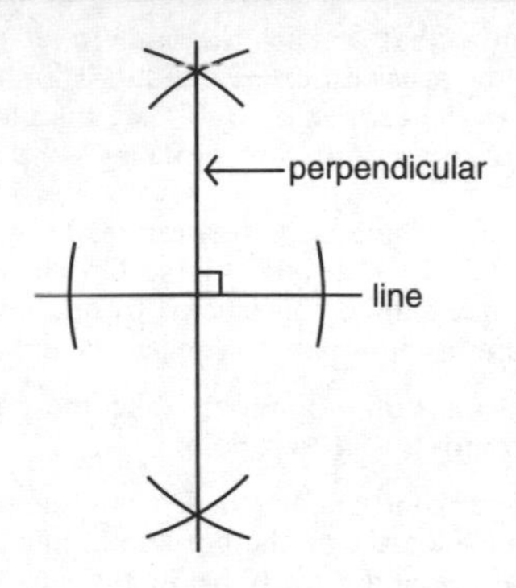

Fig. C7 *construction* (*sense 1.*)

Consumer Credit Agreement Agreement between an individual (the buyer) and a company selling goods which allows the buyer to purchase goods on credit terms, in return for the payment of **interest** on the cost of the goods. Use of **credit cards** is also covered by this type of agreement. *See* **annual percentage rate**.

continuity A function $f(x)$ is continuous at a point x_0 if $f(x) = f(x_0)$ as x tends to x_0. It is continuous in the interval $(a\ b)$ if it is continuous at all points with coordinates (a, b).

continuous variable Variable that can have all values between two stated figures. *Example*: between 2 and 3, a continuous variable could have all values between 2 and 3, such as 2·1, 2·15, 2·35, and so on.

contour Line on a map that joins places of equal altitude, i.e. places that are the same height above sea level.

convergence *1.* In geometry, the tendency of two lines to move together until they meet at a point. *2.* In a series the tendency of the sum of terms towards a definite limit. *Example*: an infinite series $a_1 + a_2 + \ldots a_n$ converges to the sum A if the sum of the terms tends to the limit A as the number of terms (n) tends to infinity (∞).

$$\lim_{n \to \infty} \sum a_n = \mathrm{A}$$

convergence of a sequence The infinite sequence $a_1 a_2 \ldots a_n$ converges to a as n tends to infinity if:

$$\lim_{A \to \infty} a_n = a$$

converse Opposite of a statement of the form 'if a is so, then b is also so'. In this case the converse is 'if b is so, then a is also so'. The converse of a true statement need not, however, also be true. For example, the statement 'if x and y are positive numbers, then $x \times y$ is always positive' is true. But the converse 'if $x \times y$ is positive, then x and y are always positive' is not true (because x and y could both be negative and still have a positive product).

conversion Changing from one number system into another. Example: (a) conversion of binary numbers to denary, 13 (denary) = 1101 (binary); (b) conversion from one base to another, $581_{10} = 2405_6$. Most **scientific calculators** will convert from and into several number bases (e.g., **hexadecimal** to **octal, sexagesimal** to **decimal**).

conversion of decimal to fraction Expressing the relevant decimal as a number divided by 10, 100 and so on.
Example:

$$0{\cdot}3561 = 0 + \frac{3}{10} + \frac{5}{100} + \frac{6}{1000} + \frac{1}{10000} = \frac{3561}{10000}$$

Most **scientific calculators** will convert from fractions to decimals, and vice-versa.

conversion of fraction to decimal Expressing the relevant fraction as the **quotient** of the divisor using **long division** where the dividend is expressed to powers of 10.
Example: $\frac{3}{16} =$

$$\begin{array}{rl} & \underline{0{\cdot}1875} \\ 16 & 3{\cdot}0000 \\ & 16 \\ & 140 \\ & 128 \\ & \ \ 120 \\ & \ \ 112 \\ & \ \ \ 80 \\ & \ \ = 0{\cdot}1875 \end{array}$$

The placing of the decimal point is determined by where in the division the dividend is expressed as a multiple of 10^1.

convex In geometry, describing a curve or surface that bulges outwards (like a dome). It occurs if no interior angle is greater than 180°.

coordinate geometry Branch of mathematics in which geometric entities such as lines, curves and solids are represented by algebraic expressions. For example, a **linear**

equation of the form $y=mx+c$ can be plotted in **Cartesian coordinates** for various values of x and y and will result in a straight line (of slope m). Other general equations of the second degree give rise to various **conic sections**, such as the circle, ellipse, parabola and hyperbola. Values of x and y that simultaneously satisfy the equations are coordinates of points that lie on the curves. Alternative name: **analytical geometry**. *See also* **polar coordinates**.

■**coordinates** Any set of numbers that fix a point in space in a certain reference frame. [3/4/e] [3/5/f] *See* **Cartesian coordinates**; **polar coordinates**.

coplanar Describing points or lines that lie in the same plane.

core Element in a computer **memory** consisting of a piece of magnetic material that can retain a permanent positive or negative electric charge until a current passes through it (when the charge changes polarity).

correct to Describing a limit on the degree of **accuracy** of a value, usually of the number of decimal places. *Example*: 8·6/3·5=2·46 correct to two decimal places, but=2·45714 correct to 5 decimal places.

■**correlation** In statistics, a relationship between two variables. Positive correlation exists when two variables increase or decrease together. Negative correlation occurs if one variable increases as the other decreases. [5/6/c]

correlation coefficient Coefficient of **correlation**, usually given the symbol r where:

$$r=\frac{\sum xy}{\sqrt{\sum x^2}\,.\,\sqrt{\sum y^2}}$$

If σ_x and σ_y are the standard deviations of x and y and n is the number of samples, then the above equation simplifies to:

$$r = \frac{\sum xy}{n\sigma_x\sigma_y}$$

If $r=0$ the two variables are completely independent. If $r=1$ there is complete correlation between them.

correspondence Two sets having the same number of elements between them. For set X and set Y this is often written as $X \rightleftharpoons Y$. These sets are also said to be equivalent. Also for n elements $n(X)=n(Y)$. However set X need not equal set Y.

■**corresponding angle** In geometry, one of a pair of angles on the same side of a **transversal** where it cuts other lines. If the other lines are parallel, the corresponding angles are equal, as in Figure C8. [4/4/c] *See also* **alternate angle**.

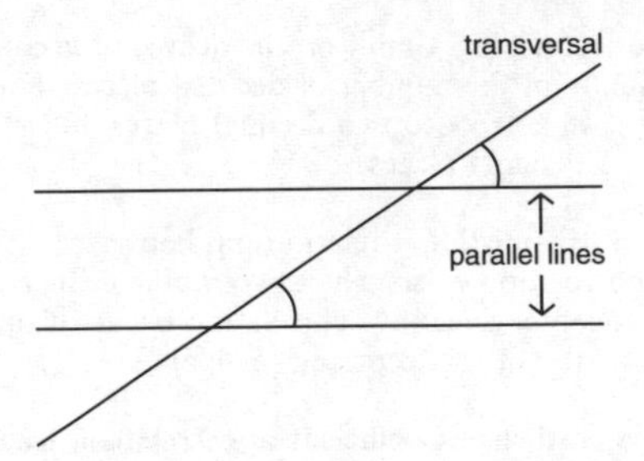

Fig. C8 *corresponding angle*

cos Abbreviation for **cosine**.

cosec Abbreviation for **cosecant**.

■ **cosecant** (cosec) **Trigonometrical ratio**. For an angle θ in a right-angled triangle, its cosecant is equal to the length of the **hypotenuse** divided by the length of the side opposite the angle, i.e. 1/**sine** θ. Tables exist for these values, or they can be found on a calculator with a 'sin' key and a **reciprocal** $\left(\frac{1}{x}\right)$ key. *Example*: if $\theta = 24°$, key 24, then 'sin', which gives 0·40673, then $\frac{1}{x}$, which gives 2·45859. [4/9/a] *See* **cotangent**; **secant**.

cosech Symbol for the **hyperbolic function** $\frac{1}{\sinh x}$. Most **scientific calculators** include a hyperbolic function key.

cosh Symbol for the **hyperbolic function** $\frac{1}{2}(e^x + e^{-x})$.

■ **cosine** (cos) **Trigonometrical ratio**. For an angle in a right-angled triangle, its cosine is equal to the length of the side adjacent to the angle divided by the length of the hypotenuse. [4/8/b] *Example*: in Figure C9, the right-angled triangle ABC is shown, with sides a, b and c. The sides AB and BC are drawn more thickly than the third side to clarify which sides of the triangle are relevant to the term cosine. The angle ABC, or $\angle$B, carries the sign θ to show that the size of the angle is not known.

From the properties of **similar triangles**, it can be shown that the **ratio** of $\frac{\text{BC}}{\text{AB}}$ is constant for $\angle$B, $= \theta$. The name given to this ratio is cosine. Using the terms a, b and c, it can also be

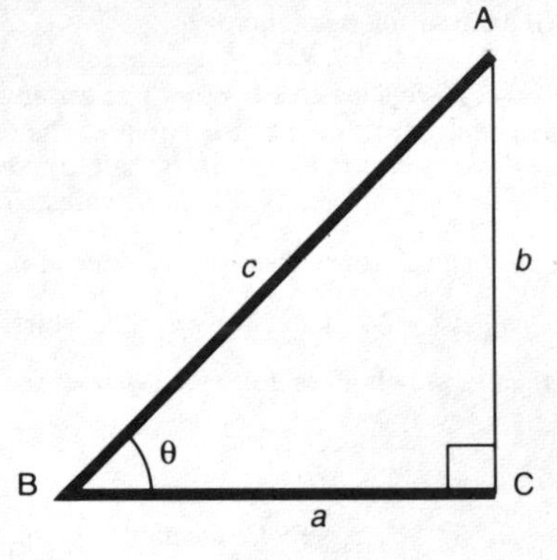

Fig. C9 *cosine*

said that

$$\cos\,\theta = \frac{a}{c}$$

$$a = c \times \cos\,\theta$$

$$c = \frac{a}{\cos\,\theta}$$

Provided that two of the variables are known in a cosine problem, the third can therefore always be found using the formulae. A further aspect of the cosine can also be considered: because triangle ABC is a right-angled triangle, it follows that the two angles other than the right angle ($\angle$A and $\angle$B) are **complementary**, i.e. they add to 90°. Using the ratio **sine**, it can be stated that the **sine** of $\angle A = \frac{a}{c}$, i.e.

$$\sin A = \frac{a}{c}$$

Cos B, however, also equals $\frac{a}{c}$. The statement can therefore be made

$$\sin A = \cos B$$

The cosine of one angle, therefore, is equal to the sine of its complementary angle. This can be stated more fully as

$$\cos \theta = \sin (90° - \theta)$$

Example: a company has estimated that the new ramp it wants to add to its loading bay should end 4 m from the loading bay wall, and lie at an angle of 45°; how long will the new ramp be? Using the triangle in Figure C9, the value 45° can be given for θ, and the value 4 m given for side a. The side we have to find, therefore, is c. Using the formula given

$$c = \frac{a}{\cos \theta}$$

Adding the values supplied gives

$$c = \frac{4}{\cos 45°}$$

First, find cos 45°, = 0·70710; now divide 4 by 0·70710 = 5·6568. Rounding up gives the solution 5·66 m. Finding the value of a cosine can be done in either of two ways: by use of a table of natural cosines, or by use of a calculator which has a 'cos' key. With a calculator, enter the number of degrees, and then press the 'cos' key. To perform the calculation in the example given above, key 4, then ÷, then 45 cos, then =. Tables of natural cosines list the cosines in the way shown below

	0′	6′	12′	18′	24′
26°	·8988	8980	8973	8965	8957
27	·8910	8902	8894	8886	8878
28	·8829	8821	8813	8805	8796

The value in degrees is given at the extreme left of the table. A full table will give cosines for 0–54 minutes (′) in intervals of 6′ (since there are 60′ in one degree of arc). Note that the **mantissa** only of the cosine is given; the **characteristic** is always 0 except for 0°, when it is 1. A calculator with a 'cos' key will give the characteristic and mantissa. Cosine tables can also be used in reverse to find the angle when the cosine is known. To do this, look for the cos value on the table, then look at the top of the column to check how many minutes of arc the cos is, and then look at the extreme left column to find the relevant value in degrees. Note that cosines (in comparison to sines and **tangents**) decrease as they approach 90°. Also, as noted above, $\cos\theta = \sin(90° - \theta)$, so the angle of any cosine can be found from a table of natural sines. *Example*: given cos 0·8746, this gives a sine value of 61°; 90° − 61° = 29°, therefore cos 0·8746 = 29° (NB this should only be used as a means of finding the approximate degree value of a cosine; always check the result on a full cosine table, or using the 'cos' key on a calculator).

■ **cosine rule** Rule based on **trigonometric ratios** which can be used in any triangle ABC with angles A, B and C and sides *a*, *b* and *c*, as in Figure C10. [4/10/b] It can be stated as follows:

(1). $$a^2 = b^2 + c^2 - 2bc \,.\, \cos \mathrm{A}$$

OR:

$$\cos \mathrm{A} = \frac{b^2 + c^2 - a^2}{2bc}$$

(2). $$b^2 = a^2 + c^2 - 2ac \,.\, \cos \mathrm{B}$$

OR:

$$\cos \mathrm{B} = \frac{a^2 + c^2 - b^2}{2ac}$$

(3). $$c^2 = a^2 + b^2 - 2ab \,.\, \cos \mathrm{C}$$

OR:

$$\cos \mathrm{C} = \frac{a^2 + b^2 - c^2}{2ab}$$

These formulae make it possible to find all or any of the angles in any triangle when the lengths of all three sides of the triangle in question are known. *Example*: construct a triangle where $a = 8$ cm, $b = 5{\cdot}6$ cm, and $c = 5{\cdot}1$ cm; find the angles using the formulae given.

$$\cos \mathrm{A} = \frac{31{\cdot}36 + 26{\cdot}01 - 64}{2(5{\cdot}6 \times 5{\cdot}1)}$$

$$= \frac{-6{\cdot}63}{57{\cdot}12}$$

$$= -0{\cdot}11607 = 96{\cdot}665$$

$$= 96^\circ\ 39'$$

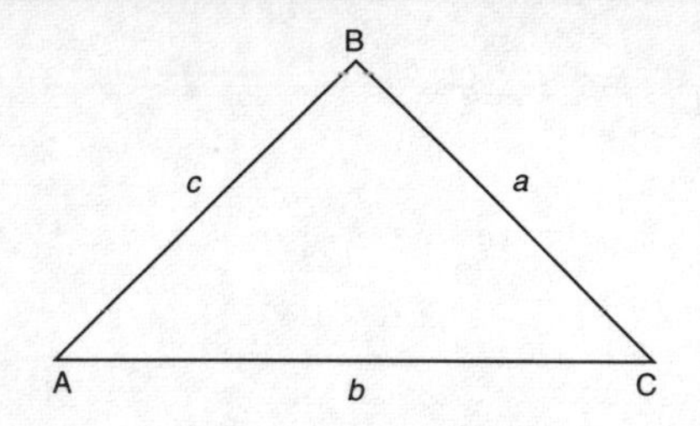

Fig. C10 *cosine rule*

$$\cos B = \frac{64 + 26{\cdot}01 - 31{\cdot}36}{2(8 \times 5{\cdot}1)}$$
$$= \frac{58{\cdot}65}{81{\cdot}6}$$
$$= 0{\cdot}7187 = 44{\cdot}04 = 44^\circ\ 2'$$
$$\cos C = \frac{64 + 31{\cdot}36 - 26{\cdot}01}{2(8 \times 5{\cdot}6)}$$
$$= \frac{69{\cdot}35}{89{\cdot}6}$$
$$= 0{\cdot}77399$$
$$= 39{\cdot}28 = 39^\circ\ 17'$$

Adding $96^\circ\ 39' + 44^\circ\ 2' + 39^\circ\ 17'$ gives $179^\circ\ 58'$, rounded up = 180°.

cot Abbreviation for **cotangent**.

■**cotangent** (cot) **Trigonometrical ratio**. For an angle θ in a right-angled triangle, its cotangent equals the length of the side adjacent to the angle divided by the length of the side opposite to it, i.e. $= 1/$**tangent** θ. Tables exist for such values, or they can be found on a calculator with a 'tan' key and a **reciprocal** $\left(\frac{1}{x}\right)$ key. *Example*: if $\theta = 65°$, key 65, then 'tan', which gives 2·14450, then $\frac{1}{x}$, which gives 0·46630. [4/8/b] *See also* **cosecant**; **secant**.

coth Symbol for the **hyperbolic function** $\frac{\cosh x}{\sinh x}$.

counter In computing, any device that accumulates totals (e.g. of repeated program loops or cards passing through a punched **card reader**).

credit card Plastic card issued by banks and credit card companies allowing the user to purchase goods or make cash withdrawals on credit. *See* **annual percentage rate**.

■**critical path analysis** Procedure used to plan complicated processes. A particular optimum sequence of events needs to take place so that the outcome is achieved in the shortest possible time. [5/10/c]

■**cross-multiplication** Method of simplifying an equation in which one or both terms are fractions. The product of the numerator on the left-hand side of the equation and the denominator on the right-hand side is equal to the product of the denominator on the left-hand side and the numerator on the other. [3/10.b] *Example*: if $3x/4 = 2y/3$, cross multiplying gives $3x \times 3 = 4 \times 2y$ or $9x = 8y$.

cross-section Plane shape that results from cutting through a solid, usually at right angles to any **axis of symmetry**. For example, a cross-section of a cylinder or cone at right-angles to the base is a circle, and a cross-section of a **tetrahedron** parallel to one of the faces is an **equilateral triangle**.

■**cube** *1.* Solid figure with six square faces. A cube has eight points and twelve edges. The unit of volume is often measured in **cubic centimetres** (cc) (which is a cube of edge length equal to 1 centimetre). *2.* Product of any number multiplied by its **square**. [4/2/a]

cube root Number which, when multiplied by itself three times, equals a given number. It is indicated by the signs $\sqrt[3]{}$ or index $\frac{1}{3}$. *Example*: the cube root of 25 ($\sqrt[3]{25}$) is 2·9240. Most **scientific calculators** have a cube root function. *See also* **square root**.

cubic centimetre (cc) Unit of volume measure on the **metric** scale.

cubic function In algebra, **polynomial** of the third degree, general formula $ax^3 + bx^2 + cx + d$.

cuboid Solid figure with six rectangular faces; opposite faces are equal. A type of **polyhedron**.

■**cumulative frequency** Frequency distribution in which the frequencies are added together to form a running total or cumulative frequency. [5/8/b] *Example*:

Marks	Frequency	Cumulative Frequency
0–10	5	5
11–20	8	13
21–30	11	24
31–40	5	29
41–50	2	31

cumulative frequency curve Frequently used method of plotting cumulative frequency data, as shown in Figure C11. Alternative term: **ogive curve**.

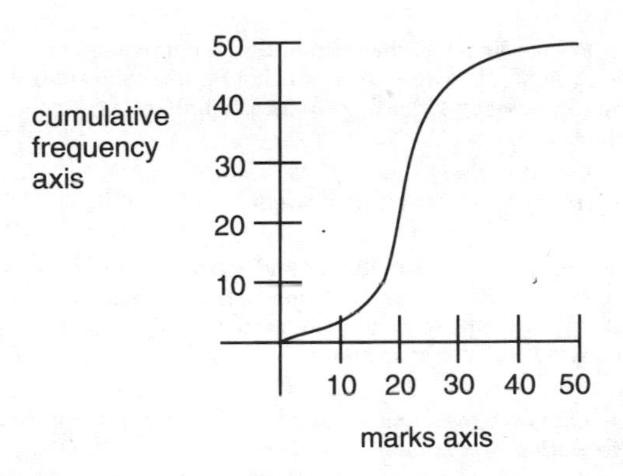

Fig. C11 *cumulative frequency curve*

currency Legal tender in a buying and selling market. The term is most often applied to the name of the legal tender of a given country, such as the pound sterling in the UK, franc in France, the yen in Japan, the escudo in Portugal, and so on. When visiting a foreign country, travellers usually buy some of that country's currency in a **bureau de change** to use during their stay abroad.

current account Account with a bank which is used for day-to-day transactions, such as cash withdrawals without notice. Current accounts are usually subject to bank charges. These are charges which the bank levies for services such as issuing chequebooks, the use of **cash cards**, and cancelling of cheques. It is also possible to arrange an **overdraft** on a current account. *Compare* **deposit account.**

cusp In **coordinate geometry**, point on a curve where it crosses itself, the two branches being on opposite sides of a common **tangent**. Often known as a double point. *See* **cycloid**.

cwt Abbreviation for **hundredweight**.

cybernetics Science that studies and attempts to build control systems resembling those of living things for mechanisms and electronic systems (e.g. in building industrial robots). Originated by Wiener in 1948.

■ **cycle** One of a repeating series of similar changes, e.g. in a **wave motion** or **vibration**, as shown in Figure C12. One cycle is equal to the period of the motion; the number of cycles per unit time is its frequency. A frequency of 1 cycle per second = 1 **hertz**. [4/9/g]

cyclic permutation Set that results from mapping each element of an ordered set to its successor (the last being mapped to the first). *Example*: $\{a,b,c\} \to \{b,c,a\} \to \{c,a,b\}$.

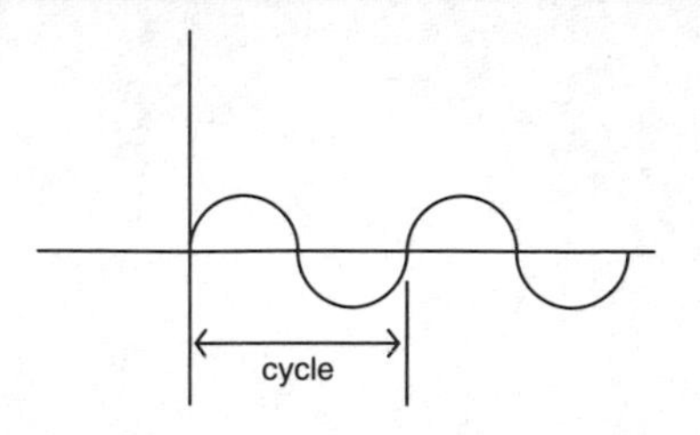

Fig. C12 *cycle*

The first of these is given by

$$\begin{pmatrix} a & b & c \\ b & c & a \end{pmatrix}$$

The degree of cyclic permutation is the number of elements in the set.

■ **cyclic quadrilateral** Four-sided figure whose vertices, or corners, lie on a circle, as shown in Figure C13. The opposite angles in a cyclic quadrilateral are supplementary (i.e. they add to 180°). In the Figure, angles ABC and ADC are supplementary. [4/6/d] The Greek astronomer Ptolemy based his work on **trigonometric ratios** on what is now known as 'Ptolemy's theorem', i.e. that in a cyclic quadrilateral ABCD,

$$AB \times CD + BC \times DA = AC \times BD$$

where AC and BD are the diagonals of the quadrilateral.

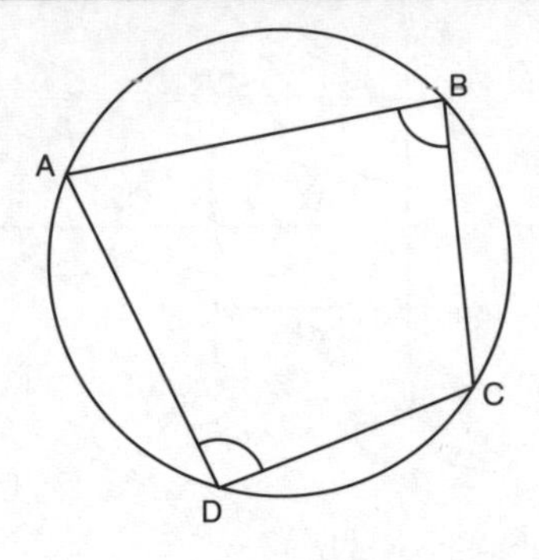

Fig. C13 *cyclic quadrilateral*

cycloid Curve traced by a point on the circumference of a circle that rolls along a straight line. *Example*: in Figure C14, a cycloid is traced from the locus of point P on the

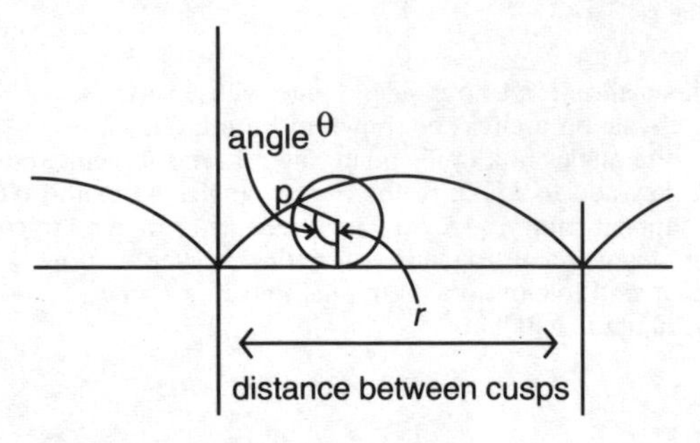

Fig. C14 *cycloid*

circumference of the small circle as it rolls along the horizontal line. The line of a cycloid can be established using the equations:

$$x = r(\theta - \sin\theta)$$

$$y = r(1 - \cos\theta)$$

The distance between the **cusps** is equal to $2\pi r$. Note that the unit of angular measure used with a cycloid is the **radian**.

■ **cylinder** Solid figure with a circular cross-section about an **axis of symmetry**. A cylinder of radius r and height h has a volume of $\pi r^2 h$ and a curved surface area of $2\pi rh$; its total surface area is $2\pi rh + 2\pi r^2$, *or* $2\pi r(h+r)$. [4/2/a]

D

D *1.* Number 13 in the **hexadecimal** number system. *2.* Number 500 in the **Roman numeral** system.

d Symbol for **deci-**, $\times 10^{-1}$.

da Symbol for **deca-**, × 10.

data Collection of information, often referring to results of a statistical study or to information supplied to, processed by or provided by a computer (excluding the **program**).

■**data bank** Alternative name for a **data base**. [5/5/a]

■**data base** *1.* Organized collection of **data** that is held on a computer where it is regularly updated and can easily be accessed (often by many users). [5/5/a] Alternative name: data bank. *2.* **Applications program** that controls and makes use of a data base.

data transmission Transfer of **data** between outstations and a central **computer** or between different computer systems.

day Unit measure of time equal to 24 **hours**. *See also* **solar day**.

debug To remove a **bug** (or fault) from a computer system.

DEC Abbreviation for **decimal notation**.

deca- (da) Prefix in the **metric** system, signifying × 10.

decagon In geometry, a ten-sided plane figure.

■**decay rate** Reduction over time of the **amplitude** of a decaying quantity. If this reduction is very rapid the decay rate is high. For example, radioactive substances decay exponentially with time (see Figure D1). [3/9/e] *See also*: **exponential series**; **half-life**.

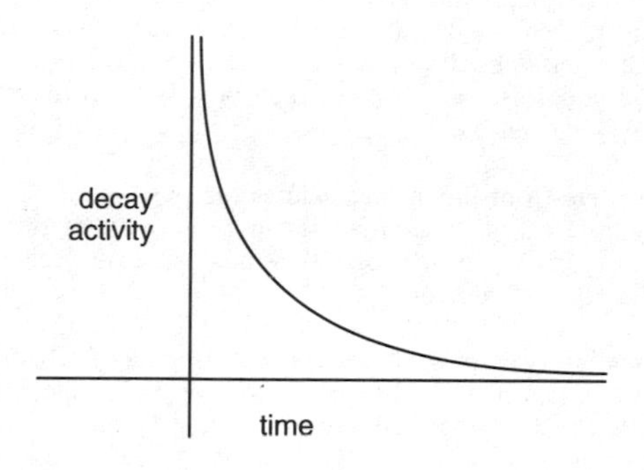

Fig. D1 *decay rate*

deci- Prefix in the **metric** system signifying $\times 10^{-1}$.

decimal notation (DEC) Manner of writing numbers using the **decimal point** and the **decimal system**.

■**decimal point** Dot written between the unit figure of a number and its fractional part when expressed in the decimal system, e.g. $31\frac{2}{10}$ is written as 31·2. [2/4/o]

■**decimal system** Number system that uses the base 10; i.e. it uses the digits 1 to 9 and 0. To convert decimals to fractions, *see* **conversion of decimal to fraction**. *See also* **binary notation**; **denary system**. [2/4/1]

■**decision tree diagram** Visual means of representing a series of events for which there are two possible outcomes at each stage. [5/4/e] [5/8/e] [5/9/d]. *Example*: as shown in Figure D2, six red balls and six yellow balls are placed in two boxes A and B. A has 3 red and 1 yellow ball and B has 3 red and 5 yellow balls. Each box is selected by tossing a coin and then one ball is selected from each box. What is the chance of getting a red ball?

declination Coordinate that defines the position of a heavenly body in terms of its angular distance along the meridian North (positive declination) or South (negative declination) of the **celestial equator**.

■**definite integral Integral** that has exact limits. [3/10/c] The definite integral of the function $f(x)$ over the range $x=a$ to $x=b$ is the area enclosed from the x axis to the curve $y=f(x)$ between $x=a$ and $x=b$. *Example*: the definite integral of x^2 between $x=3$ and $x=4$ is:

$$\int_3^4 x^2 \, . \, dx = \left[\frac{x^3}{3}\right]_3^4 = \left(\frac{4^3}{3} - \frac{3^3}{3}\right) = \left(\frac{64}{3} - \frac{27}{3}\right) = 21\tfrac{2}{3} - 9 = 12\tfrac{2}{3}$$

■**degree** *1.* In geometry, the unit derived by dividing a circle into 360 segments, used to measure angles and describe direction; it is subdivided into minutes and seconds (of arc). Its symbol is °. [4/4/c] *Compare* **radian**. *2.* In algebra, the **power** or index of the **variable**: x^2 is of the 2nd degree.

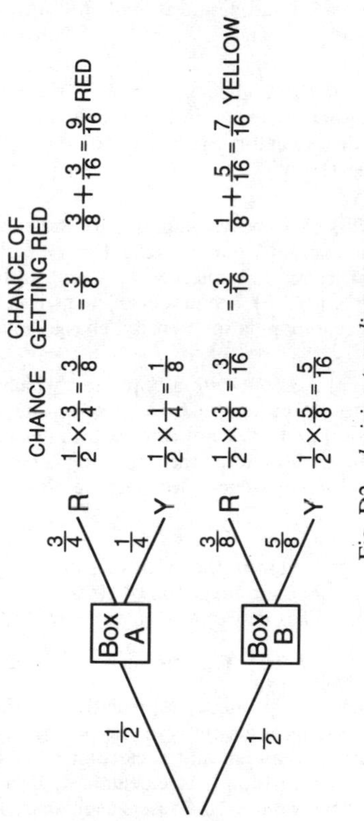

Fig. D2 *decision tree diagram*

denary system Number system using the base 10. Its integers are 0,1,2,3,4,5,6,7,8 and 9, and it is often known as the **decimal system**.

■**denominator** Lower number of a **fraction** (the upper number is the **numerator**). *Example*: in the fractions 2/3, 7/12 and 83/100, the denominators are 3, 12 and 100. [2/4/k] *See also* **common denominator**.

■**density** *1.* Mass of a unit **volume** of a substance. For an object of mass m and volume V, the density d is m/V. It is commonly expressed in units such as g/cm^3 (although the SI unit is kg/m^3). [2/7/h] *2.* Number of items in a defined surface area (e.g. population density, charge density).

deposit account Account with a bank into which money is put subject to conditions, such as limits on the amount of any withdrawal and prior notice of withdrawal. In return for observing these restrictions, the depositor is paid **interest** on the money. *Compare* **current account**; *see also* **savings account**.

depreciation Reduction in the value of something over time. For example, a house bought for £42,000 and sold two years later for £36,000 has depreciated in value by £6,000, or $14{\cdot}3\% \left(= \frac{6{,}000}{42{,}000} \times 100\right)$. The term can also be applied to the financial assets in a business. Depreciation on these has to be included in the **annual** accounts of the business as an expense of the business and an amount is calculated for a given time period. The usual method is to calculate depreciation as a percentage of the value of the assets each year. This is called the reducing balance method.

derivative In mathematics, the result of the **differentiation** of a mathematical **function**, equivalent to the rate of change of a function or the slope of a curve at a particular point. If $y=f(x)$, the derivative is written as

$$\frac{dy}{dx}$$

or $f'(x)$. The derivative of $y=a/b$ where a and b are functions of x is:

$$\frac{dy}{dx}=\frac{(\mathrm{V}du/dx-\mathrm{U}dv/dx)}{\mathrm{V}^2}$$

See also **calculus**.

desktop publishing (DTP) Technique that uses a **PC** linked to a **word processor** (with access to various type founts and justification programs) and a laser **printer** to produce multiple copies of a document that rivals conventional printing in quality. The addition of a **scanner** allows the introduction of simple graphics (illustrations).

determinant In mathematics, a quantity obtained by adding products of elements of a square **matrix** according to certain rules. If there are n rows and n columns, then n^2 is the number of elements or the order of the determinant.

■**diagonal** Line joining any two non-adjacent corners of a **polygon**. A polygon with n sides has $\frac{1}{2}n(n-3)$ diagonals. [4/4/a]

■**diameter** Length of a straight line that bisects a circle; it passes through the centre of the circle and is equal to twice the **radius** ($2r$) of the circle. Diameter measurement is one of the principal determinants in calculating the volume of any circular **solid**, such as a **cylinder**. *Example*: the diameter of a circle with a radius of 5 cm = 10 cm; the volume of a cylinder with the same diameter, and a height of 4 cm = $2\pi r^2 h = 2\pi \times (10/2)^2 \times 4 = 628{\cdot}3185$ **cubic centimetres**. [4/2/a]

difference Number remaining after subtraction has taken place. If the difference is zero then this is equality.

difference of (or between) two squares Way of clarifying expressions describing the difference between squared powers by means of **factorisation**. *Example*: in the expression $x^2 - y^2$ (which is the difference between two squares) the result can be factorized into $(x-y)(x+y)$.

differential calculus Branch of mathematics that deals with continuously varying quantities. It uses **differentiation** for calculating such values as rates of change, slopes of curves, and maximum and **minimum** values.

differentiation Method used in **calculus** to determine the **derivative** f' of a **function** f. If $f(x) = Ax^n$, $f'(x) = nAx^{n-1}$. If $y = f(x)$ is the equation of a curve (as in **coordinate geometry**), $f'(x)$, or dy/dx, is the slope of the curve at the point x.

differentiator **Analog computer** device whose (variable) output is proportional to the time differential of the (variable) input.

digit In mathematics and computing, each of the numerals 0–9 inclusive is considered a digit.

digital Describing numerical representations of values, as opposed to **analog** representations. For example, on a analog clock with hour and minute hands and a second sweep one might be able to read the time as 4·20 and 10 seconds; on a digital clock this might read as 16:21:12.

digital/analog converter Device that converts digital signals into continuously variable electrical signals for use by an **analog computer**.

digital computer Computer that operates on **data** supplied and stored in digital or number form (as opposed to continuously variable data supplied to an **analog** computer).

digital display Display that registers values as digits, as in a pocket **calculator**, for example.

dimension *1.* Power to which a fundamental unit is raised in a derived unit. *Example*: **acceleration** has the dimensions $[L/T^{-2}]$, i.e. +1 for length (L) and −2 for time (T), equivalent to length divided by the square of time. *2.* Length, height and width of a **solid**. All three are separate dimensions; taken together they describe the overall dimensions of a three-dimensional object.

dimensional analysis Prediction of the relationship of quantities. If an equation is correct the **dimensions** of the quantities on each side must be identical. It is an important way of checking the validity of an equation.

direct proportion Increase or decrease of two variables at the same rate. Alternative term: **direct variation**.

direct variation *See* **direct proportion**.

■**directed numbers** Either positive or negative numbers. They

appear on either side of zero, e.g. +8, −5. [2/3/i] There are certain rules for different operations:
Addition:

$$(+)+(+)=(+)$$

$(+)+(-)=$: take the smaller number from the larger one. The sign will be that of the larger one.

Subtraction: Change the sign of the negative number and add the result.

$$(+)-(-)=(+)$$

Multiplication:

$$(+)\times(+)=(+)$$
$$(-)\times(-)=(+)$$
$$(-)\times(+)=(-)$$
$$(+)\times(-)=(-)$$

Division:

$$(+)/(+)=(+)$$
$$(-)/(+)=(-)$$
$$(+)/(-)=(-)$$
$$(-)/(-)=(+)$$

directrix Line that determines the shape of a **conic section**. *See also* **eccentricity**.

discount Amount by which the selling price of an article is reduced. It is usually a certain percentage of the selling price, and can often be a reduction for a cash sale. *Example*: If a desk with a marked selling price of £60·00 is sold for cash at a discount of 10%, how much does the cash purchaser pay? The customer pays:

$$\pounds(60{\cdot}00-(10/100\times 60))=\pounds60{\cdot}00-\pounds6{\cdot}00$$
$$=\pounds54{\cdot}00$$

Using a calculator, key $60-(60\times 10\%)=$.

discrete In mathematics, that which is not broken up into smaller values.

discrete distribution Frequency distribution which is made up of **integer** values. For instance this could be the number of heads if 3 coins were tossed a number of times. The distribution can only be frequencies with 0, 1, 2 or 3 heads, not parts of them. A distribution of this typc is usually represented by a frequency line histogram, not a bar histogram.

discrete variable Variable that can only have an **integer** value, e.g. 0, 1, 2.

discriminant Quantity b^2-4ac, derived from the coefficients of a **quadratic equation** of general formula $ax^2+bx+c=0$.

disjoint set Set of numbers that does not overlap or intersect another set of numbers, e.g. the set of even numbers and the set of odd numbers are disjoint.

disk Magnetic disc used to record data in computers. *See also* **floppy disk, hard disk**.

diskette Alternative name for a **floppy disk**.

dispersion Scatter of a set of data. It is often measured in statistics by the **standard deviation**.

displacement Position of one point relative to another, including both the distance between the two points and the direction of the first point from the second point. A **vector** quantity.

display Short name for a **liquid-crystal display** (LCD) or a **visual display unit** (VDU).

distance In mathematics, the length of a straight line joining two points.

■**distribution** Arrangement of a set of elements in a **set**. A normal distribution is a symmetrical distribution which shows the data of a normal frequency curve. A skewed distribution is a non-symmetrical distribution. A symmetrical distribution is one which is symmetrical or the same, about the **mean**. [5/10/a] *See* **frequency distribution**.

distributive law Link between the mathematical operations of multiplication and addition. $x \times (y+z) = (xy) + (xz)$. However, in Boolean algebra $x + (yz) = (x+y)(x+z)$. In set theory:

$$x \cup (y \cap z) = (x \cup y) \cap (x \cup z)$$

where $\cup$ = union and $\cap$ = intersect.

diverge To move apart. *1.* In sequences, a divergent sequence is one that does not converge to a finite limit. *2.* In series, a divergent series is a series that does not converge to a finite limit. *3.* In vectors, the divergence of a vector is the **scalar product** of $\nabla.\mathbf{v}$ of a vector $\mathbf{v}$ with the vector operator del(∇). It can be written as div.**v**.

dividend In mathematics, a number to be divided by another one (the **divisor**). The result of the division is the **quotient**. *Example*: in 64/18, 64 is the dividend, and 18 is the divisor. The quotient is $3{\cdot}\dot{5}$.

division Inverse of multiplication, in which a dividend x is divided by a divisor y to give a **quotient** z: $x/y=z$.

■ **division of fractions** To divide a number by a fraction, the number is multiplied by the inverted fraction. [2/5/c] *Example*: $\frac{3}{8}$ divided by $\frac{9}{16}$ is equal to $\frac{3}{8}\times\frac{16}{9}=\frac{2}{3}$. **Scientific calculators** with a fraction ($a\frac{b}{c}$) key will divide fractions.

divisor Number that is divided into another one (the **dividend**). The result of the division is the **quotient**.

dodecagon Plane figure, or **polygon**, with 12 sides. The sum of the interior angles of a dodecagon is equal to 1800°.

dodecahedron Solid figure, or polyhedron, with 12 faces. In a regular dodecahedron, the faces are regular **pentagons**.

domain Set of points which is often called a **region**. The domain of a mathematical function is the set of values of the independent variable for the particular function for which it is defined. *Example*: The domain of a set of ordered pairs of numbers:

$$\{(0,0)\ (2,1)\ (6,3)\ (12,6)\}$$

is the first element of the pairs:

$$0,\ 2,\ 6,\ 12$$

dominance matrix In a square matrix $n\times n$, a row a is said to dominate (or be larger than) another row b if every number in row *a* is as large or larger than the corresponding entry in row *b*. Similarly, one column can dominate another column for the same reason.

domino rule Term used to describe events happening in sequence one after the other, each event happening as a consequence of the preceding event.

DTP Abbreviation of **desktop publishing**.

duodecimal system Number system that uses the **base** 12, requiring two numbers, usually called A (or *T*) and B (or *E*), in addition to 1 to 9 in order to represent 10 and 11 respectively.

dynamics Branch of **mechanics** that deals with the actions of forces on objects in motion. It can be divided into statics (which deals with objects at rest), and kinetics (dealing with the effect of forces on the motion of objects).

E

E Number 11 in the **duodecimal** number system.

E *1.* Number 14 in the **hexadecimal** number system. *2.* Symbol for exa-, $\times 10^{18}$. *3.* Abbreviation for **compass** direction East.

e *1.* Fundamental mathematical constant and the base of natural (Napierian) **logarithms**. It is an **irrational number**, the limiting value of the series $(1+1/n)^n$ as n tends to infinity, approximately equal to 2·71828. It is the constant in an **exponential function** or **series**. *2.* The symbol used for the **coefficient of restitution** or for the **eccentricity** of a **conic section**.

eccentric Not having the same centre, as in a circle drawn around a previous circle, with a different centre.

eccentricity (e) *1.* In mathematics, property of a **conic section** (ellipse, parabola or hyperbola) equal to the distance from the curve to a fixed point (**focus**) divided by the distance to a fixed line (**directrix**). For the three curves listed, it is less than 1, equal to 1 and greater than 1, respectively. The eccentricity of a circle is 0. *2.* In astronomy, the extent to which an orbit (of a planet or satellite) departs from a circle.

ecliptic Apparent path of the Sun on the **celestial sphere**, which passes through the 12 constellations of the Zodiac. It intersects the **celestial equator** at the two **equinoxes**.

■**edge** *1.* Boundary line of a surface. *2.* Line where two surfaces of a **solid** meet. For example a **cube** has 12 edges. [1/2/b]

element *1.* One of the members of a **set**. The symbol $\in$ is used to signify when a value is part of a set, as in $2 \in \{1, 2, 3, 4\}$. When a value is not part of a set, the symbol $\notin$ is used, as in $5 \notin \{1, 2, 3, 4\}$. *2.* A value in a **matrix**. *3.* In geometry, a line, plane or point.

elevation *1.* Vertical distance of a point from the horizontal. *2.* Vertical **plane** of a **solid**. In architectural drawings, for example, the front of a building will be termed the front elevation, the rear of the building the rear elevation, and so on. *See also* **face**; **depression**.

■ **elimination** Joining of two simultaneous equations to reduce them to one containing only one variable instead of two. In the process, the other variable is eliminated from the equation. [3/7/g] *Example*: solve the equations

$$\text{(a)} \quad 3x + 2y = 12$$

$$\text{(b)} \quad x + 5y = 17$$

Multiplying (b) by 3 to make the coefficients of x in both equations the same gives

$$\text{(c)} \; 3x + 15y = 51$$

Eliminating x by subtracting (a) from (c) gives

$$\begin{array}{r} 3x + 15y = 51 \\ -\,3x + 2y = 12 \\ \hline 13y = 39 \\ y = 3 \end{array}$$

To find x we substitute 3 for y in either of the two original equations:

$$3x+2\times3=12$$
$$3x=12-6$$
$$3x=6$$
$$x=2$$

Hence the solutions are: $x=2$; $y=3$. *See also* **substitution**.

ellipse *1.* Plane curve, a **conic section** generated when a cone is cut by a plane non-parallel to the base. *2.* The **locus** of a point that moves round two fixed points, called **foci**, in such a way that the sum of its distances from the foci remains constant. It has an **eccentricity** of less than 1. *See also* **Kepler's laws**.

ellipsoid Solid figure of which all plane sections are **ellipses**.

■ **empirical probability** Probability based on previous experience or experiment. 'Empirical' is the opposite of 'theoretical': it describes concrete action. *Example*: empirical proof of the probability of throwing a six using a die is found by actually throwing the die a certain number of times, taking notes of the results, and then calculating the probability using the notes obtained. [5/5/i]

empty set Set containing no elements, symbol $\varnothing$.

■ **enlargement Transformation** that multiplies all sides of a figure by a simple number or scale factor greater than 1. The final shape is similar to the original. [4/7/e] *See* **centre of enlargement**.

envelope In geometry, a curve that is tangential to every curve of a whole **family of curves**.

epicycle Curve traced by the centre of a circle that rolls round the outside of the circumference of another larger circle. The **parametric equations** are

$$x = (a+b)\cos\theta + b\cos\left[\frac{(a+b)\theta}{b}\right]$$

$$y = (a+b)\sin\theta - b\sin\left[\frac{(a+b)\theta}{b}\right]$$

where a = the radius of the fixed circle, b = the radius of the rolling circle and θ = the angle between the line of centres of the rolling sphere and the x axis. **Cusps** occur when the curve and the fixed circle touch each other. The number of cusps = a/b.

epicycloid Curve traced by a point on the circumference of a circle that rolls round the outside of the circumference of another circle. If both circles have the same diameter, the epicycloid is a **cardioid**. Using the formula used in the definition of **epicycle**, this happens when $a/b = 1$.

equal *1.* In mathematics, similarity or likeness in a relationship or property between two quantities, symbol =. *2.* In geometry, similarity or likeness in a relationship or property between two parts of a geometric figure. Alternative term: **congruence**.

■ **equality of matrices** Equality of the corresponding elements of matrices to each other. [4/10/d] *Example*:

$$\begin{pmatrix} k & l \\ m & n \end{pmatrix} = \begin{pmatrix} p & q \\ r & s \end{pmatrix}$$

$k = p$, $l = q$, $m = r$, $n = s$.

■**equally likely outcomes** In probability theory, the equal likelihood of two or more results. [5/6/f] *Example*: a tossed coin is equally likely to fall head up as tail up. In probability a certain event has a probability of 1. Something that can never happen has a probability of 0. Something which is equally likely to happen to something else has a probability of 1/2.

■**equation** Mathematical statement of **equality** which often contains unknown algebraic quantities such as x and y. [3/5/e]

■**equations of motion** Equations for the motion of an object moving in a straight line. [3/9/d] Five **parameters** can describe this type of motion:

initial velocity	v_1
final velocity	v_2
acceleration	a
distance travelled	s
time taken	t

Thcsc give rise to five equations, each containing only four of the parameters.

$$s=\tfrac{1}{2}t(v_1+v_2)$$

$$s=v_1t+\tfrac{1}{2}at^2$$

$$s=v_2t-\tfrac{1}{2}at^2$$

$$v_2=v_1+at$$

$$v_2{}^2=v_1{}^2+2as$$

equator, terrestrial Imaginary **great circle** that divides the

Earth into the Northern and Southern Hemispheres. It runs round the greatest circumference of the Earth and is equally distant from the two poles. *See also* **celestial equator**.

equatorial *1.* Relating to the **equator** (terrestrial). *2.* In mathematics, describing a plane figure in which all the sides are equal.

equidistant Of any two points, equally distant from a third point. *Example*: any two points on the **circumference** of a **circle** are equidistant from the **centre** of that circle.

■ **equilateral** Having equal sides, e.g. an **equilateral triangle** has all its sides equal. [4/2/a]

equilateral triangle Regular triangle where all the sides are equal in length and all the angles are equal in size. [4/2/a]

equilibrium Zero sum or resultant of all external forces acting on a body. There are three types of equilibrium: stable, unstable and neutral. *Example*: in Figure E1, a cone is shown in the three positions stable, unstable and neutral.

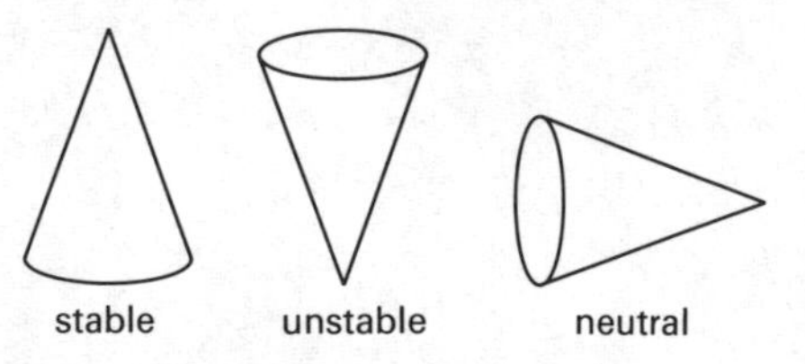

Fig. E1 *equilibrium*

In the stable position, after a slight displacement the cone will return to its original position. In the unstable position, a slight displacement will cause the cone to topple over and not resume its original position. In the neutral position, a slight displacement will cause the cone to remain in its displaced position.

equinox Moment when the Sun appears to cross the **celestial equator**, and day and night are of equal length. There are two equinoxes each year: the vernal (or spring) equinox and the autumnal equinox. In the northern hemisphere, the vernal equinox is approximately 21st and 22nd of March and the autumnal approximately 22nd or 23rd of September.

equivalence *1.* Mathematical properties that are equal. *2.* Mathematical properties that belong to the same **set**.

equivalent fractions Two fractions show **equivalence** if the ratio of the **numerator** to the **denominator** is the same in both cases, e.g. 3/5 is equivalent to 9/15.

error Margin of difference between an **estimate** and a correct value. There are different ways of describing by how much or how little the error departs from the correct value. For example, it is possible to say 'within ±6 cm' or, in the case of decimals, 'to 3 decimal places'.

■**estimate** Guess or approximation about the answer to a numerical problem. It should be fairly close to the actual answer when it is finally calculated. [2/4/p] *See also* **absolute error**.

Euclidean geometry Describing ordinary three-dimensional geometry, named after the Greek mathematician Euclid (*fl.*

300 BC). He set out five principles, or **axioms**, which formed the basis of **geometry** in the West for almost 2,000 years. They are:

1. a straight line can be drawn from one point to any other point;
2. any straight line can be continued infinitely in the same straight line;
3. a circle can be described with any centre and any radius;
4. all right angles are equal to one another;
5. if a straight line meets two other straight lines so as to make the sum of the two interior angles on one side of the transversal less than two right angles, the other straight lines will, if extended infinitely, meet on that side of the transversal.

Axiom 5 (known as the *parallel postulate*) leads to particular mathematical problems when a proof is attempted and, as a way of resolving this difficulty other, non-Euclidean, geometries were created in the 19th century. *See also* Appendix III.

Euler's formula *1.* In geometry, the formula which states that the number of **vertices** plus the number of **faces** minus the number of edges of any **polyhedron** equal 2. This can be written as

$$V+F-E=2$$

where V = vertices, F = faces and E = edges. *Example*: in a **cube** there are 8 vertices, 6 faces and 12 edges, and $8+6-12=2$.
2. Formula for **exponential functions** which states: $e^{ix} = \cos x + i \,.\, \sin x$.

even number Whole number that is not an odd number and has no remainder when divided by 2.

■ **event** In set theory, an outcome. It could be **discrete** or continuous. [5/4/g,h,i]

evolute Curve traced by joining the centres of curvature of points on another curve (called the **involute**).

exa- (E) Prefix in the **metric** system signifying $\times 10^{18}$.

exchange rate Rate at which sums in one **currency** can be valued in terms of another. *Example*: a traveller wants to change £150·00 into German deutschmarks at a time when the rate is 2·775 marks to the pound. The traveller will therefore receive $150{\cdot}00 \times 2{\cdot}775 = 416{\cdot}25$ marks. Exchange rates vary over time, so it is always a good idea to check (in a newspaper, bank, or **bureau de change**) what is the current rate for a particular currency.

exp Abbreviation for **exponent**.

expansion Change in the form of a function of a mathematical quantity into some form of power series. *Example*: for a geometric progression the general term ax^{r-1} can be expanded to form the series $a + ax + ax^2 + \ldots ax^{n-2} + ax^{n-1}$.

■ **expected probability** The likely probability of something happening. [5/4/g,h,i] Alternative term: **empirical probability**.

■ **experimental error** Difference between the actual or true value of an expression and that found by investigation. It may be due to experimental technique, the accuracy of measuring instruments, or uncontrollable factors such as human error or arbitrary data. [2/10/a]

exponent (exp) Number that indicates the power to which a quantity (the **base**) is to be raised, usually written as a superior number or symbol after the quantity, e.g. in 2^3 and 2^x, the exponents are 3 and x, and the bases are 2 in both cases. Most **scientific calculators** include an x^y function which allows any number x to be found to the power y.

exponential function Function e^x. The term e forms the base of **natural** or **Napierian logarithms**. *See also* **e**; **Euler's formula**.

exponential series Mathematical **series** of functions of x that converges to e^x. *Example*:

$$1+x+\frac{x^2}{2!}+\frac{x^3}{3!}+\ldots\frac{x^{n-1}}{(n-1)!}$$

When $x=1$

$$e=1+1+1/2!+1/3!+\ldots$$

exterior angle Angle formed between one side of a **polygon** and the extension of the adjacent side.

extrapolation Estimation of a value outside the range of those already known, usually by graphical methods. *Example*: in Figure E2, the line of pressure is extrapolated backwards to arrive at the value for the absolute zero of temperature (−273 °C).

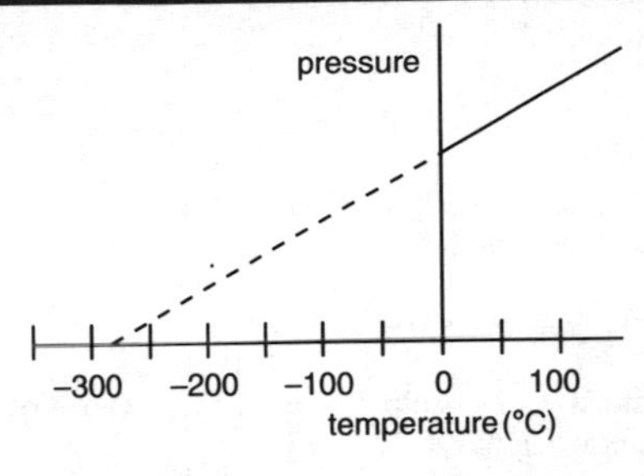

Fig. E2 *extrapolation*

F

F *1.* Number 15 in the **hexadecimal** number system. *2.* Symbol for **Fahrenheit**.

f Symbol for **femto-**, $\times 10^{-15}$.

■**face** Flat surface of a **solid** figure. *See also* **elevation**; **Euler's formula** (sense *1.*); **facet**. [4/2/b]

facet Flat side, or face, of a many-sided object, such as a cube or a cut gemstone.

■**factor** *1.* Number that will divide exactly into another given number (e.g 2 and 3 are factors of 6). *2.* **Polynomial** that will divide exactly into another polynomial (e.g. $(x+2)$ and $(x-5)$ are factors of $x^2-3x-10$). [3/5/c]

factor theorem Theorem which shows that a number or expression is the outcome of two or more factors multiplied together. *Example*: a^2-b^2 is the outcome of the factors $a+b$ and $a-b$ being multiplied together.

factorial Product of all the whole numbers from a given whole number n down to 1, written as $n!$. *Example*: the factorial of 6 (written 6!) is:

$$6\times5\times4\times3\times2\times1=720.$$

The factorial of $3=6$, of $4=24$, of $5=120$. Factorials quickly become large numbers; $10!=3{,}628{,}800$. Most **scientific calculators** have a factorial function key.

■**factorization** Splitting up of a number or algebraic expression into the parts that form the original number or expression when multiplied together. [3/10/b] *Example*:

$$\text{(a)}\quad 6x+6y=6(x+y)$$

The factors of $6x+6y$ are therefore 6 and $x+y$.

$$\text{(b)}\quad 24x^2-6x=6x(4x-1)$$

The factors are therefore $6x$ and $4x-1$.

factorize To split up a number or algebraic expression into its factors. If no factors can be formed then the number or expression is prime. *See* **prime number**; **sieve**.

Fahrenheit scale Temperature scale on which the freezing point of water is 32 °F and the boiling point 212 °F. It was named after the German physicist Gabriel Fahrenheit (1686–1736). This scale has now largely been superseded by the **Celsius scale**. To convert Fahrenheit temperatures to Celsius temperatures, subtract 32 and multiply by $\frac{5}{9}$. *Example*: $75\ °\text{F}=75-32=43\times 5=215/9=23\frac{8}{9}=23{\cdot}8\ °\text{C}$. *See* Appendix III.

family of curves Set of curves that display a particular geometric pattern. *Example*: in Figure F1, $y=mx+c$ when $c=0, 1, 2, 3 \ldots$ gives a family of curves that, in this case, are straight lines.

Farey sequence Series of fractions of the form a/b in their lowest terms in order of increasing size (or magnitude). With $1\leq a\leq n$ and $1\leq b\leq n$ a Farey sequence of order n could be obtained. *Example*:

order 3: $\frac{1}{3}\ \frac{1}{2}\ \frac{2}{3}\ 1$

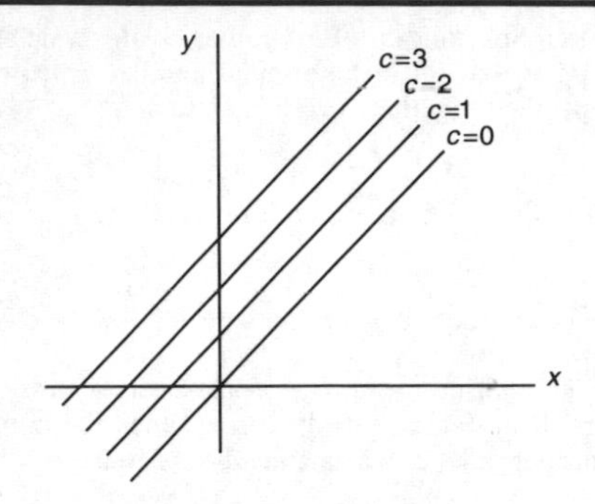

Fig. F1 *family of curves*

order 5: $\frac{1}{5}\ \frac{1}{4}\ \frac{1}{3}\ \frac{2}{5}\ \frac{1}{2}\ \frac{3}{5}\ \frac{2}{3}\ \frac{3}{4}\ \frac{4}{5}\ 1$

order 7: $\frac{1}{7}\ \frac{1}{6}\ \frac{1}{5}\ \frac{1}{4}\ \frac{2}{7}\ \frac{1}{3}\ \frac{2}{5}\ \frac{3}{7}\ \frac{1}{2}\ \frac{4}{7}\ \frac{3}{5}\ \frac{2}{3}\ \frac{5}{7}\ \frac{3}{4}\ \frac{4}{5}\ \frac{5}{6}\ \frac{6}{7}\ 1$

a/b, c/d form a Farey sequence if $bc-ad=1$.

fathom *1.* **British unit** used to express depth of water, particularly at sea. It is equal to 6 feet. *2.* In mining and the timber trade, a fathom is a volume of 6 cubic feet.

feet Plural of **foot**.

femto- (f) Prefix in the **metric** system signifying $\times 10^{-15}$.

Fibonacci series Sequence of whole numbers 1, 1, 2, 3, 5, 8, 13, 21, 34, 55, . . . , in which each number is the sum of the two preceding numbers. It was named after the Italian mathematician Leonardo Fibonacci (*c.* 1175–*c.* 1240). *See also* **golden section.**

field In computing, specific part of a **record**, or a group of characters that make up one piece of information.

figure *1.* Character representing a number, such as any of the **Arabic numerals** 1 to 9 inclusive. *2.* Combination of lines and points in geometry: a **triangle** is a geometric figure.

finite Having a precise limit. *Compare* **infinite.**

finite series Series with a precise limit. *Example*: 1, 2, 3, . . . $n-1$, n is a finite series.

floppy disk Flexible, portable magnetic disk that provides data and program storage for **PC**s. The disk may be enclosed in a flexible or a rigid casing. Alternative name: diskette. *See also* **hard disk.**

■**flowchart** Visual way of presenting a sequence of events or operations using 'boxes' of different shapes to distinguish stages in the sequence. [5/7/e] *Example*: in Figure F2, a flowchart is used to present the sequence 6 multiplied by 4 divided by 2.

fluid ounce Unit of liquid measure, abbreviated as fl oz, equal to 28·41 **cubic centimetres** (cm^3).

foci Plural of **focus.**

focus Point on a **conic section** where two lines of every pair of conjugate lines passing through it are perpendicular to each other (see Figure F3). The **ellipse** and **hyperbola** each

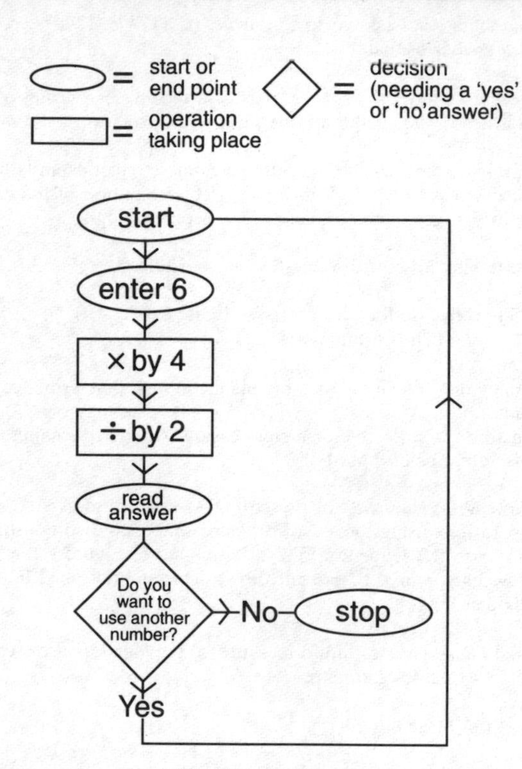

Fig. F2 *flowchart*

have two real and two imaginary **foci**. In Figure F3, if A is any point on the section and F is the focus, and B is the base of the perpendicular from A to the directrix, $AF = eAB$ where e is the eccentricity. $e < 1$ for an ellipse, $e = 1$ for a parabola, and $e > 1$ for a hyperbola.

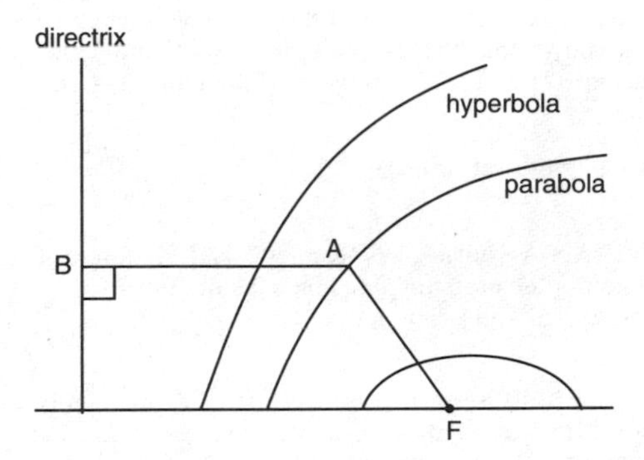

Fig. F3 *focus*

■**foot** Unit of length in the **f.p.s.** system, plural feet. It contains 12 equal parts called inches and is equal to one-third of a **yard** = 0·3048 metres. Abbreviation ft, symbol ′. [2/5/h]

force Influence that can make a stationary object move, or a moving object change speed and/or direction, i.e., that changes the body's momentum. For a body of mass m and acceleration a, the force $F = ma$.

■**formula** (plural formulae) Mathematical expression that shows a constant relationship between various quantities; e.g. the formula for the area A of a circle of radius r is $A = \pi r^2$. The area of a circle, therefore, is always arrived at by applying this sequence of operations, whatever the quantities of the **variables** (i.e. radius and radius squared in this case). Also, having the answer to a formula makes it possible to find the variables concerned if they are unknown. *Example*: if it is known that the area of a circle is 127 mm^2, then the radius must be $\sqrt{(127/\pi)}$ mm = 6·35809 mm. [3/4/d]

formulae Plural of **formula**.

FORTRAN Acronym of FORmula TRANslation, a high-level computer programming language designed for mathematical and scientific use.

Fourier analysis Mathematical method of expressing a complex function that represents a wave as a series of simpler sine waves. It was named after the French mathematical physicist Jean Fourier (1768–1830). It is an infinite trigonometrical series of the form

$$\begin{aligned} f(x) = \tfrac{1}{2}a_o &+ (a_1 \cos x + b_1 \sin x) \\ &+ (a_2 \cos 2x + b_2 \sin 2x) + \ldots \\ &+ a_n (\cos nx + b \sin nx) + \ldots \end{aligned}$$

where the coefficients are constants.

f.p.s. system System of measurement based on the **foot**, **pound** and **second**. See also **c.g.s. system**.

■**fraction** Part of a whole, represented mathematically by a pair of numbers. The upper numerator is written above the lower denominator and separated from it by a horizontal or diagonal line; e.g., $\frac{2}{3}$, $\frac{7}{10}$, 3/7. Fractions of a hundred may be given as percentages; e.g. $\frac{65}{100}=65$ per cent (sometimes written 65%). In decimal fractions, the denominators are powers of 10 (10, 100, 1,000, . . .), usually written using a decimal point and place values as for whole numbers. Thus $\frac{7}{10}=0{\cdot}7$, and $\frac{27}{1000}=0{\cdot}027$. [2/2/f] Most **scientific calculators** have a key which allows values to be entered as fractions. To convert a fraction with a denominator which is not decimal to one which is, *see* **conversion of fraction to decimal**. *See also* **common denominator**.

frame of reference Set of points, lines or planes for defining positions.

■**frequency** Number of times a certain event occurs in a particular unit of time. [2/5/i]

■**frequency distribution** Way of organising raw data into an ordered form, usually a list showing the number of times each event occurs, hence the name frequency distribution. See **cumulative frequency** and **cumulative frequency curve**. [5/10/b]

frustum Solid formed by cutting through another solid with two parallel planes; e.g. a conical frustum (shaped like a lampshade or bucket), where two parallel lines cut a cone. Often referred to as a **truncated** solid.

ft Abbreviation of **foot** or **feet**.

function Mathematical expression used in **analysis**, made up of one or more **variables** and numbers and constants. The function of a single variable such as x is often written as $f(x)$. *Example*: if $x=4$, then in the function $f{:}x \rightarrow 3x+1$, $f=$ 13, and the function maps 4 to 13; if $f{:}x \rightarrow 2x^2-3$, and x is still 4, then $f=29$, and the function maps 4 to 29. *See also* **derivative**; **differentiation**; **Taylor series**.

furlong Unit measurement of length on the **f.p.s. system**. One furlong = 220 **yards** = 201·168 metres. Eight furlongs equal one **mile**.

G

G Symbol for giga-, $\times 10^9$.

g Abbreviation of **gram**.

gal Abbreviation of **gallon**.

■**gallon** (gal) Unit of liquid volume or capacity equal to eight pints (on the British standard). One gallon of pure water at 62 °F and 30 inches of atmospheric pressure has a mass of 10 pounds; 1 gallon =4·547 litres=4 **quarts**. [2/5/h]

gate In computing, an electronic circuit (switch) that produces a single output signal from two or more input signals. Alternative name: logic element. *See also* **logic circuit**.

geometric mean Of n positive numbers or quantities, the nth **root** of their overall **product**. *Example*: the geometric mean of 3 and 4 is the second (square) root of 12, which is 3·464. *See also* **arithmetic mean**.

geometric progression (G.P.) Series in which each term is obtained from the previous term by multiplying or dividing it by a constant called the common ratio. e.g. in the series 1,2,4,8, . . . the common ratio is 2. The general expression for a geometric progression for n terms is:

$$a, ar, ar^2, \ldots ar^{n-1}$$

The sum of the G.P. is equal to:

$$a(1-r^n)/(1-r)$$

When n approaches infinity, the sum tends to $a/(1-r)$.

geometry Branch of mathematics that is concerned with the properties of curves, surfaces and solids in space. *See also* **Euclidean geometry**.

giga- (G) Prefix denoting $\times 10^9$, one thousand million. A gigavolt (GV), for example, is 10^9 volts.

glide reflection Transformation which occurs when a **translation** is followed by a **reflection**.

golden mean *See* **golden section**.

golden section Ratio between two parts of a line of fixed length such that the ratio of the smaller part to the larger part equals the ratio of the larger part to the whole: i.e. a line AB divided at point C gives the ratio

$$AC : CB = AB : AC$$

The ratio can be expressed numerically as 1·61803 or $\frac{1}{2}(1+\sqrt{5})$. It was known to the ancient Greeks, and since then has acquired the association of being a particularly satisfying ratio for use in constructing proportions in painting, sculpture and architecture. The ratio also corresponds to the terms of a **Fibonacci series**, and can be found in the natural world, where it is evident in the growth patterns of, for example, plants and shells. *Example*: one **construction** for the golden section is as follows:

1. take any line of fixed length. Construct a square ABCD on the line;

2. select the mid-point of the base of the square, calling it P; draw a diagonal from that point to the upper right vertex of the square;
3. using the length of the diagonal as **radius**, draw an arc of a circle using P as the centre of the circle;
4. produce the base of the square to meet the arc drawn, calling the point where the two meet F;
5. draw a right angle to CF at F and produce AB to meet the right angle at E. In the rectangle so formed, $\frac{DF}{CD}=\frac{CD}{CF}$.

goniometer Instrument for measuring angles, particularly the angles between the faces of crystals.

G.P. Abbreviation of **geometric progression.**

grad Abbreviation for **grade.**

grade Unit employed in measuring **angles**. One grade $=0{\cdot}9^\circ=\frac{1}{100}$ of a right angle.

■**gradient** Amount of inclination of a line (or a curve at a particular point) to the horizontal; its slope. The gradient of a curve at a given point is the slope of the tangent to the curve at that point. [3/9/f] *Example*: for a curve $y=f(x)$ the gradient can be found by calculating $f'(x)$ the first derivative (dy/dx). If a line is drawn between the points (x_2, y_2) and (x_1, y_1) then the gradient is

$$(y_2-y_1)/(x_2-x_1)$$

■**gram (gramme)** (g) Unit of mass on the **c.g.s. system**, equal to 1/1000 kg = 0·0352 **ounces**. [2/5/i]

■**graph** Mathematical diagram with two axes, each representing the change in value of a single variable, on which the user visually plots the movement of the two variables in relation to each other. [3/8/f] *See also* **bar chart**; **histogram**; **pie chart**.

graphical calculator Electronic **calculator** that has (along with a wide range of other functions) an inbuilt facility for creating and displaying **graphs** using **Cartesian** (x and y) coordinates, and **bar charts**. Such calculators will create graphs using specified values for x and y, and will also display pre-programmed graphs for standard **trigonometric ratios** such as **sine**, **cosine** and **tangent**, and for **hyperbolic functions** such as sinh, cosh and tanh.

great circle *1.* Circle whose centre is also the centre of a sphere. *2.* Imaginary line on the Earth's surface that divides the globe in half; e.g. the **equator**. The shortest distance between any two points on Earth follows the great circle that joins them.

greater than Describing the difference between values in an approximate way, symbol $>$. *Example*: $10 > 3$ means '10 is greater than 3'.

gross Amount before expected deductions are made. *Example*: when a **wage** or **salary** is quoted in a job advertisement, this will be the gross figure before deductions are made for items such as income tax and National Insurance contributions. *Compare* **net**.

group Any collection of data.

■**grouped data** In statistics, any data collected together for a particular purpose. [5/2/b]

■**growth rate** The increase of a quantity over time. *Example*: £300 invested at 6·5% **simple interest** over 10 years will have a growth rate of 6·5% per year. *See also* **compound interest**. [3/9/e]

H

h Symbol for hecto-, $\times 10^2$.

ha Abbreviation for **hectare**.

■**half-life** Time taken for the activity of a radioactive substance to fall to half its original value, as shown in Figure H1, where H = half-life and N = activity. [3/9/e]

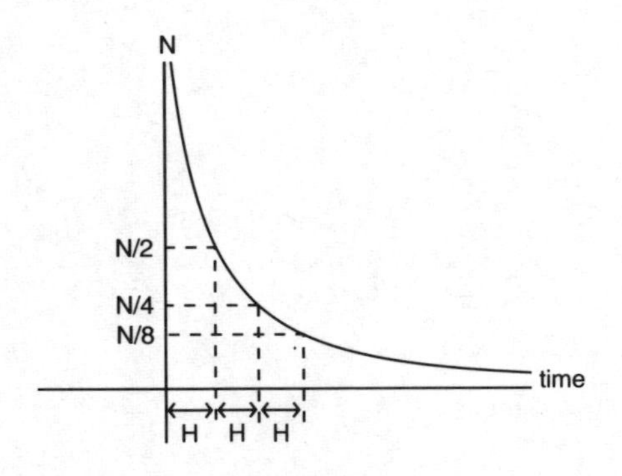

Fig. H1 *half-life*

half line Line extended in one direction only.

hard disk Rigid magnetic disk that provides data and program storage for computers, including **PCs**. Hard disks can hold a high density of data. When used in a PC, they are usually not portable. *See also* **floppy disk**, **Winchester disk**.

hardware Electronic, electrical, magnetic, and mechanical parts that make up a **computer** system. *See also* **software**.

harmonic progression Series of numbers whose reciprocals have a constant **difference** (i.e. form an **arithmetic progression**), e.g. $1/1+1/2+1/3+1/4$.

HCF Abbreviation for **highest common factor**.

head Electromagnetic component in a tape recorder, video recorder, record player or computer input/output device that can read, erase or write signals off or onto tapes and disks.

■**hectare** (ha) Unit of **area** on the **metric system**, equal to 100 **ares**. [2/5/i] Related measures are as follows:

1 are $= 100\ \text{m}^2$

1 hectare $= 2{\cdot}47$ acres

100 hectares $= 1\ \text{km}^2 = 0{\cdot}386\ \text{miles}^2$

hecto- Prefix in the **metric** system signifying $\times 10^2$.

helix Curve that cuts the surface of a cylinder or cone at a constant angle, e.g. the shape of the thread of a bolt or screw. It can be represented by the **parametric equation**:

$$x = r\,.\cos\theta,\ y = r\,.\sin\theta,\ z = (r\theta)\,.\cot\alpha$$

where θ = the angle of revolution, r = the radius of the cylinder and α the angle of inclination of the helix.

■**hemisphere** Half a sphere. The sphere is cut on a plane through its centre. [4/2/a]

heptagon Plane figure, or **polygon**, with seven sides; the seven interior angles add to 900°.

Hero's formula Formula that gives the **area** of a triangle of sides a, b, and c, as follows:

$$\sqrt{s(s-a)(s-b)(s-c)}$$

where $s=\frac{(a+b+c)}{2}$. It was named after the Greek scientist and mathematician Hero of Alexandria (*fl. c.* 62 AD). *See* Appendix III.

HEX Abbreviation for **hexadecimal**.

hexadecimal (HEX) Describing a number system based on 16, commonly used in digital **computers**. Its digits are 1–9, A, B, C, D, E, F. Most **scientific calculators** will perform calculations in hexadecimal.

■**hexagon** Plane figure with six sides. The interior angles of a regular hexagon add to 720° and the length of each diagonal equals twice that of a side. [4/2/a] *See* **regular polygon**.

highest common factor (HCF) Largest number that divides exactly into each of two or more numbers, e.g. the HCF of 21, 35 and 63 is 7.

■**histogram** Type of **bar chart** in which quantities are proportional to the areas of the bars rather than to their lengths. For example, if the widths (or intervals) are equal, then the height of the columns (or the lengths) are proportional to the frequencies (see Figure H2). [5/9/a]

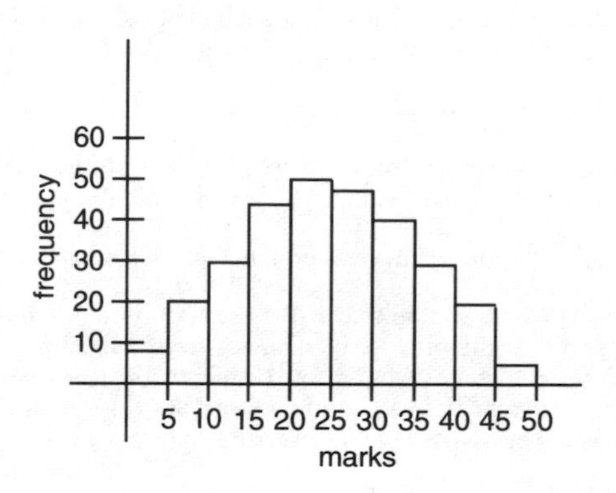

Fig. H2 *histogram*

homogeneous In mathematics, describing a **polynomial** whose terms all have the same degree. *Example*:

$$2x^3 - 3x^2y + xy^2 + 5y^3$$

is homogeneous (of degree 3).

Hooke's Law In physics, the law which states that the load applied to an elastic material is proportional to the extension, provided the elastic limit is not exceeded. The elastic limit is the point at which Hooke's Law ceases to be obeyed. *See also* Appendix III.

horizontal *1.* Line parallel to the Earth's surface. *2.* Line at right angles to the vertical.

hour (hr) Unit of time measurement; one hour = 60 **minutes** = 3,600 **seconds**.

hr Abbreviation of **hour**.

hundredweight (cwt) Unit measure of weight on the **f.p.s. system**; one hundredweight = 112 **pounds** = 50·80 kg.

hyp Abbreviation for **hyperbolic function**; **hypotenuse**.

hyperbola Conic section that is the locus of a point which moves so that the ratio of its distance from a fixed point (the **focus**) to its distance from a fixed straight line (the **directrix**) is always greater than 1; i.e. its **eccentricity** (e) is greater than one. The general equations for a hyperbola are:

$$x^2/a^2 - y^2/b^2 = 1;$$

and

$$b^2 = a^2\,(e^2 - 1)$$

hyperbolic function (hyp) Function related to a regular **hyperbola** in the way that the **trigonometric functions** (sin, cosine, etc.) are to a circle. The hyperbolic functions are: sinh, cosh, tanh, cosech, sech and coth:

$$\sinh x = 1/2(e^x - e^{-x}),$$

$$\cosh x = 1/2(e^x + e^{-x})$$

$$\tanh x = \sinh x/\cosh x$$

$$\coth x = \cosh x/\sinh x$$

$$\operatorname{sech} x = 1/\cosh x$$

$$\operatorname{cosech} x = 1/\sinh x$$

hypocycloid Curve traced by a point on the circumference of a circle that rolls round the inside of another (larger) circle. A hypocycloid has a series of **cusps**. *Example*: in Figure H3

$$x=(a-b)\,.\,\cos\theta+b\,.\,\cos\left[\frac{(a-b)\theta}{b}\right]$$

$$y=(a-b)\,.\,\sin\theta-b\,.\,\sin\left[\frac{(a-b)\theta}{b}\right]$$

where a is the radius of the fixed circle, b is the radius of the rolling circle and θ is the angle between the line of centre and the x axis. *See also* **cycloid**.

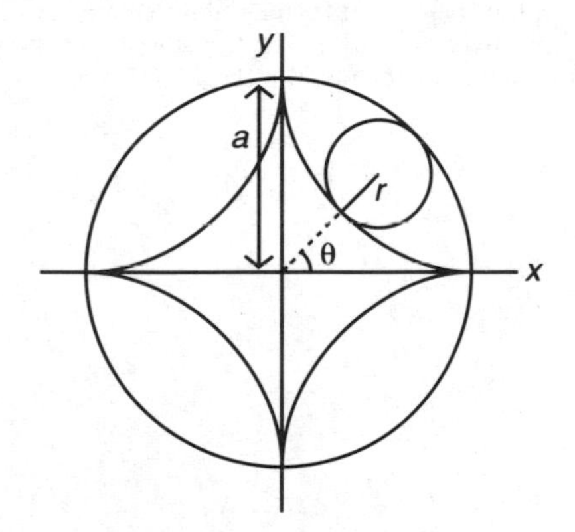

Fig. H3 *hypocycloid*

■**hypotenuse** (hyp) *1.* Side of a **right-angled triangle** that can form a square equal to the sum of the squares of the lengths of the other two sides. *2.* Longest side of a right-angled triangle. *3.* Side opposite the right angle in a right-angled triangle. [4/7/c] [4/9/a] *See* **Pythagoras, theorem of**; **trigonometrical ratio**.

hypotheses Plural of **hypothesis**.

■**hypothesis** (plural hypotheses) Idea or assumption that can be tested to find out whether or not it is accurate. *Example*: a person observes a small pool of water in a room. If the water is near a central heating radiator, the person may form the hypothesis that the water has come from the radiator, having first found out whether the water could have come from another source (did anyone bring water recently into the room; did they spill it; is the ceiling above the water wet; has it rained recently?). The hypothesis can then be tested by examining the radiator. [1/6/c]

I

I *1.* Symbol for **interest**. *2.* Number 1 in the **Roman numeral** system

i Symbol for the **square root** of −1 ($\sqrt{-1}$). *See* **imaginary number**.

icosahedron Solid figure or **polyhedron** with 20 faces. The faces of a regular icosahedron are **equilateral triangles**.

identity element In set theory, for set A there is an identity element i where $i*a=a*i=a$ for all values of a where $*$ is any binary operation. See also **unit matrix**.

identity matrix Matrix containing a single **identity element**.

■ **image** Reflection of a point at a point, or from a line, or from a plane. [4/3/b] *Example*: In Figure I1, the distance from point P to the x axis = the distance from the image of P to the x axis.

imaginary number Product of a **real number** and i, where $i^2 = -1$. It corresponds to a **complex number** $(x+iy)$ in which the real part (x) is 0.

imaginary part Non-real part of a **complex number**; i.e. in the complex number $x+iy$, iy is the imaginary part. *See also* **imaginary number**.

implicit function Variable x is an implicit function of y when x and y are connected by a function that is not explicit (i.e. in which x is not directly expressed in terms of y).

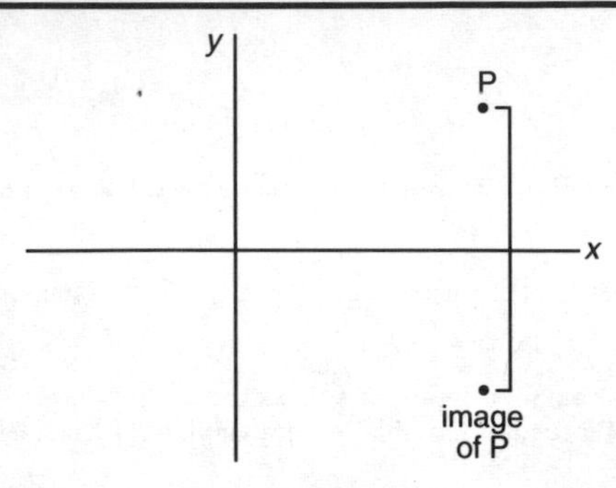

Fig. I1 *image*

improper fraction Fraction whose upper part (numerator) is larger than the lower part (denominator); e.g. $\frac{7}{3}$, $\frac{11}{4}$, $3\frac{2}{25}$. It is always greater than 1, as can be seen by converting it to a **mixed number** (the previous examples become $2\frac{1}{3}$, $2\frac{3}{4}$, $1\frac{7}{25}$).

in Abbreviation for **inch.**

■**inch** (in) Twelfth part of a **foot** in the **f.p.s. system**. 1 inch = 25·4 mm. [2/5/h]

incircle Circle that touches all three sides of a triangle inside the triangle itself. The centre of the circle is where the bisectors of the angles of the triangle intersect. Alternative term: **inscribed circle.**

■**independent event** Event that occurs in isolation without any interaction with anything else. Nothing else influences it. It may, however, go on to influence other events. [5/4/g,h,i]

independent variable If y is a function of x, i.e. $y=f(x)$, x is the independent variable of the function (and y is the dependent variable).

index In mathematics, an **exponent** or power; e.g. in the terms 5^4 and $x^{\frac{1}{2}}$, 4 and $\frac{1}{2}$ are **indices**. Most **scientific calculators** have an index key (x^y).

■**indices (laws of)** Rules in mathematics governing terms with powers or indices. [2/5/k] [2/8/b] [3/7/c] The rules can be summarized as follows:

Multiplication: when powers of the same quantity are multiplied add the indices, e.g.:

$$(a^5)(a^4)=a^{5+4}=a^9$$

Division: when powers of the same quantity are divided, subtract the index of the denominator (bottom part) from that of the numerator (top part):

$$\frac{a^5}{a^4}=a^{5-4}=a$$

Powers: when raising the power of one quantity multiply the indices:

$$(a^5)^4=a^{5\times 4}=a^{20}$$

Negative indices: these are written as the **reciprocal** of the quantity:

$$a^{-1}=1/a$$

Fractional indices: here the denominator is the root to which the quantity must be taken and the numerator is the power to which it must be raised:

$$a^{2/3} = \sqrt[3]{a^2}$$

Zero index: any quantity raised to the power zero is 1:

$$a^0 = 1$$

■ **inequality** Mathematical statement in which one quantity is greater or less than the other. Symbols for these are as follows:

$x > y$ means x is greater than y

$x < y$ means x is less than y

$x \geq y$ means x is greater than or equal to y

$x \leq y$ means x is less than or equal to y

These inequalities are not affected if the same quantity is added to, or subtracted from, both sides, or each side is multiplied or divided by the same number. However, multiplication or division by a negative number does change the inequality. [3/7/b] [3/8/g]

inequation Equation in which one side is greater or less in value than the other. *Example*: $7x + 2 > 0$, or $x - 4 \geq 3$.

infinite Without any precise limit, i.e. potentially going on without end. Opposite of **finite**.

infinite set **Set** in which it is not possible to name all the members of the set, e.g. the set of odd numbers.

infinity *1.* Quantity that is larger than any quantified concept. It may be considered as the **reciprocal** of zero.

Symbol ∞. *2.* Quantity that, when any number is subtracted from it, remains the same.

inflection Turning point on a curve where

$$\frac{dy}{dx}=0$$

and

$$\frac{d^2y}{dx^2}=0$$

As the value of x increases the sign of the gradient goes from + to 0 to + or from − to 0 to −. *Example*: see Figure I2.

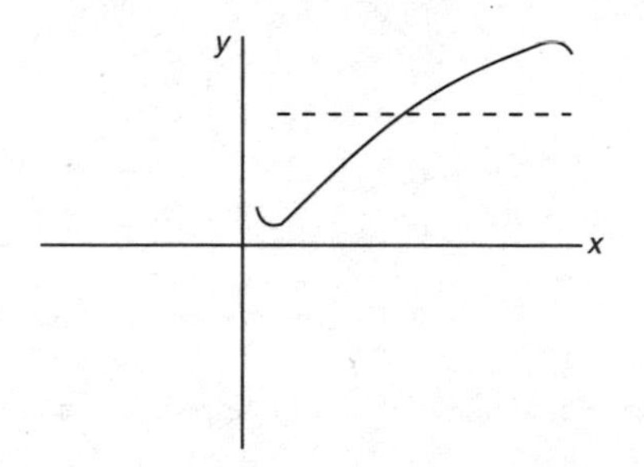

Fig. I2 *inflection*

information retrieval Science of storing and accessing data, which may use microfilm, microfiche, magnetic tape and computer storage devices.

input device Part of a **computer** that feeds it with **data** and **program** instructions. The many types of input devices include a keyboard, punched-card reader, paper-tape reader, optical character recognition, light pen (with a VDU) and various types of devices equipped with a **read head** to input magnetically recorded data (e.g. on magnetic disk, tape or drum).

inscribed circle Circle drawn inside a triangle that touches all three sides of the triangle; its centre is where the bisectors of the three angles meet, as in Figure I3, where the centre of the inscribed circle is D. Alternative name: **incircle**.

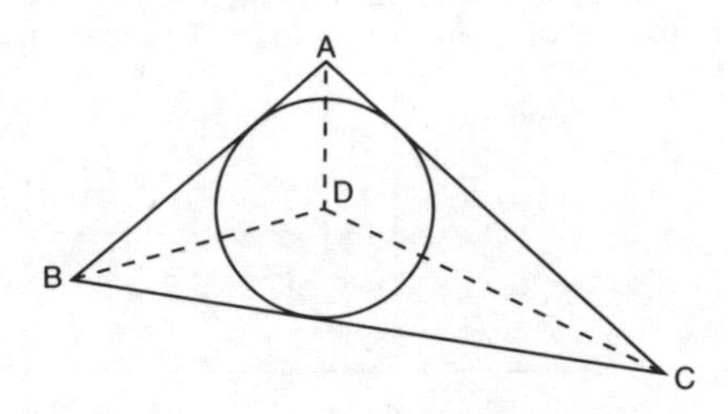

Fig. I3 *inscribed circle*

integer Whole number; it may be positive or negative.

■**integral** Value that results from the process of integration. [3/10/c]

■**integral calculus** Branch of **calculus** concerned with the process of summation (integration) of the series of infinitely small quantities (infinitesimals) that theoretically make up

the difference between two values of a given function. It is the inverse of **differentiation** and is used in the solution of such problems as finding the area enclosed by a given curve or the volume enclosed by a given surface. [3/10/c]
Example:

$$\int x^4 . dx = \frac{x^{4+1}}{4+1} + C = \frac{x^5}{5} + C$$

where C is the constant of integration. For a definite integral:

$$\int_2^4 x^4 . dx = \left[\frac{x^5}{5}\right]_2^4$$

$$= \frac{4^5}{5} - \frac{2^5}{5}$$

$$= \frac{1024}{5} - \frac{32}{5}$$

$$= \frac{992}{5} = 198\tfrac{2}{5}$$

intercept *1.* Point at which a part of a curve or plane is cut by any other curve or plane. *2.* Point at which a straight line graph cuts the y-axis.

interest (I) Money paid to a financial **investor** or charged to a financial borrower. The rate of interest is usually expressed as an annual percentage of the sum invested or borrowed. Interest is commonly paid to investors who, for example, hold **savings accounts**, such as bank **deposit accounts**. It is usually charged on financial loans, such as **Consumer Credit Agreements**, **credit card** accounts, bank **overdrafts**, and **mortgages**. *See also* **annual percentage rate**; **compound interest**; **joint variation**; **simple interest**.

interface In computers, **hardware** (and **software** to drive it) that connects one piece of equipment to another.

interior angle Angle formed by any two **adjacent** sides of a **polygon**.

interpolation Estimation of the value of a function $f(x)$ for a value of the variable x that lies between those for which the function is known.

■ **interquartile range** In statistics, a measure of the spread of the sample data, i.e. the difference between one quarter and three quarters of a sample. [5/8/c] *See* **quartile**; **box-and-whisker plot**.

intersection *1.* Common parts of curves or surfaces, which may be points, curves, or parts of a plane. *2.* In set theory, the combination of two sets that have common members which form a **subset**. This intersection of sets can be shown on a **Venn diagram**. *Example*: if set X and set Y each contain some common members, they can be represented by a Venn diagram as shown in Figure I4, where the shaded area is the intersection. This can be written using the set notation $X \cap Y$ which means X intersection Y.

interval Set of points or real numbers between two specified points or real numbers. Intervals can be of three kinds: open, closed, or mixed. A closed interval is one where the end points are known. *Example*:

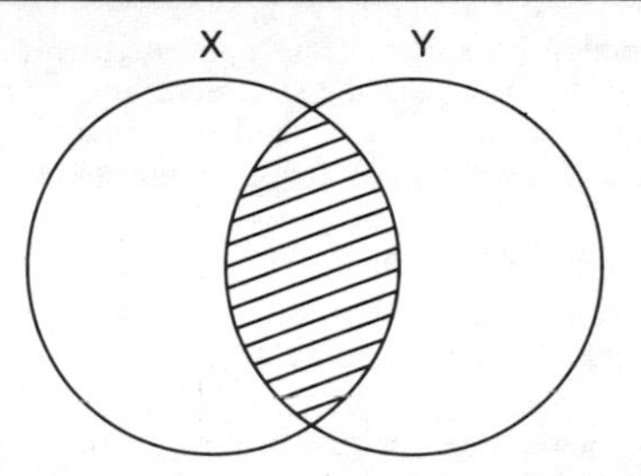

Fig. I4 *intersection (sense 2.)*

(a) if a and b are the end points of a set of numbers of any value x, $a \leq x \leq b$ is a closed interval;

(b) an open interval could be represented as $a < x < b$;

(c) a mixed interval could be represented as $a \leq x < b$ or $a < x \leq b$.

invariant Mathematical property that remains constant under certain **transformations**.

■ **inverse** Reciprocal of a number. If x is a number, $\frac{1}{x}$ or x^{-1} is the inverse. If x is inversely proportional to y then this can be written as $x \propto 1/y$, or $y \propto 1/x$. Most **scientific calculators** have an inverse (reciprocal) key, marked $\frac{1}{x}$ or x^{-1}. Many also have an inverse function key, marked INV, which finds the inverse of particular functions (e.g. $\cos^{-1}$ from cos). [3/7/d]

inverse function Function that is opposite in effect or nature to a specific function, i.e. the starting point of one function is the conclusion of the other, and vice-versa; e.g. **inverse hyperbolic functions**, and **inverse trigonometrical functions**.

inverse hyperbolic function Inverse of the respective **hyperbolic functions**: $\sinh^{-1}$ (arc-sinh), $\cosh^{-1}$ (arc-cosh), $\tanh^{-1}$ (arc-tanh), cosech^{-1}, sech^{-1} and $\coth^{-1}$.

inverse trigonometrical function Inverse of the respective **trigonometrical functions**: $\sin^{-1}$ (**arc-sine**), $\cos^{-1}$ (**arc-cosine**), $\tan^{-1}$ (**arc-tangent**), cosec^{-1}, $\sec^{-1}$ and $\cot^{-1}$.

investor Person who deposits money with a financial body for an agreed return, such as an annual rate of **interest**. *See* **annual percentage rate**.

involute Curve traced by the end of a string as it is kept tight while being unwound from another curve (e.g. a cylindrical spool that is not free to rotate). *See also* **evolute**.

■**irrational number** Real number that cannot be expressed as a fraction, i.e. as the ratio of two integers; e.g. e, π, $\surd$. [2/9/a] *See also* **surd**.

■**isometric** *1.* Describing axes radiating from a common point with 120° between each of them. *2.* Describing a **transformation**, such as a **reflection** or **translation**, when all lengths remain the same. [4/6/a]

isomorphic Properties which have a one-to-one connection, e.g. isosceles and equilateral triangles are isomorphic because they each have three sides.

isomorphism Property of having one-to-one correspondence.

isosceles triangle Triangle that has two angles and two sides equal, as in Figure I5.

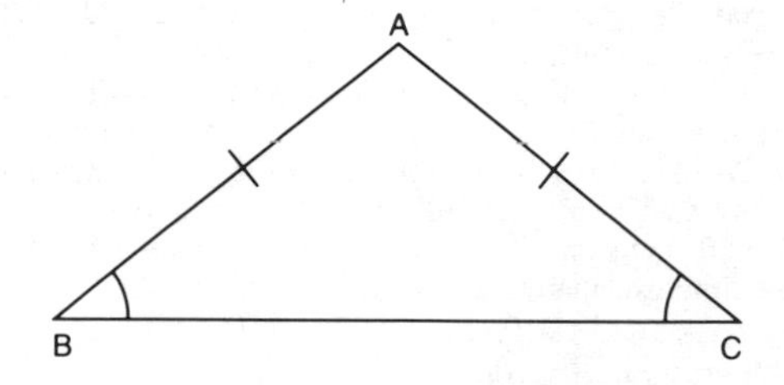

Fig. I5 *isosceles triangle*

iterate To repeat a process.

iterative Describing repetition of a subroutine in a **computer** program, e.g. when an **algorithm** is repeated.

J, K

joint variation Use in a formula of more than two **variables**. For example: it is possible to have a **direct variation** between one of the variables and the product of the others. This can be seen in the formula for **simple interest** ($I = PRT/100$), where there are four variables, and a change in any of three (P, R & T) will result in a change in the fourth. *Example*: £250 invested at 6% per annum over 3 years gives interest (I) of £45·00; the same sum at the same rate over 5 years = £75·00 interest, therefore I varies directly as T.

K Abbreviation of **kelvin**.

k Symbol for **kilo-**, $\times 10^3$.

kelvin SI unit of temperature.

Kepler's laws Series of laws governing the movement of planetary bodies first described by the German astronomer and mathematician Johannes Kepler (1571 – 1630). They can be summarized as follows:

1. Planets orbit in ellipses, with the Sun at one focus of the ellipse.
2. If a line is drawn from the Sun to a planet, it traces an equal area in any equal periods of time.
3. Squares of the times of orbit of the planets are proportional to the cubes of their average distances from the Sun.

See Appendix III.

keyboard Computer **input device** consisting of a standard **qwerty** keyboard, usually with additional function keys, which a human operator uses to type in **data** as **alphanumeric** characters.

kg Abbreviation of **kilogram**.

kilo- (k) Prefix meaning 1,000 ($\times 10^3$).

■ **kilogram** (kg) Unit of mass in the **S.I. units** system = 1,000 g = 2·2046 **pounds**. [2/5/i]

kilometre (km) Unit of length in the **S.I. units** system, = 1,000 m = 0·621 **miles**. [2/5/i]

■ **kinematics** Branch of **mechanics** concerned with interactions between **velocities** and **accelerations** of various parts of moving systems. [2/7/h]

kite In geometry, a four-sided figure in which the diagonals intersect at right angles and the two pairs of adjacent sides are equal.

km Abbreviation for **kilometre**.

km/h Abbreviation of kilometres per hour.

kn Abbreviation for **knot**.

■ **knot** (kn) Unit of speed used in navigation and meteorology. 1 knot = 1 nautical mile per hour = approximately 1·15 land miles per hour, = 1·85 km/h. [2/5/h]

Königsberg bridge problem Example of a 'classic' mathematics problem, set by the Swiss mathematician Euler (1707 1783). In the town of Königsberg (called Kaliningrad since 1945, when it became part of the then USSR) there were seven bridges, positioned as shown in Figure K1. Euler's question was: can a person cross all the seven bridges at Königsberg once in a single journey? In Figure K1, the bridges are shown by the lines a, b, c, d, e, f, g. All attempts to solve the problem have failed to find a solution. Euler used the topological graph shown in Figure K2 to tackle the problem. The lines in the graph represented the bridges. He denoted a point on the graph as odd if an odd number of segments ended at the point, and a point as even if an even number of segments did. He then made three statements:

(a) if ALL points are even the journey is in fact possible and the person finishes at the starting point.

(b) if one or two points are odd the journey is possible but the person ends at a different point from where they started.

(c) if more than two points are odd, the journey is not possible.

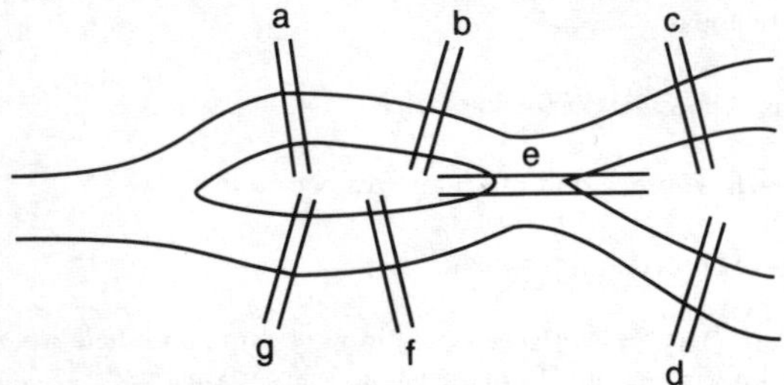

Fig. K1 *Königsberg bridge problem (bridges)*

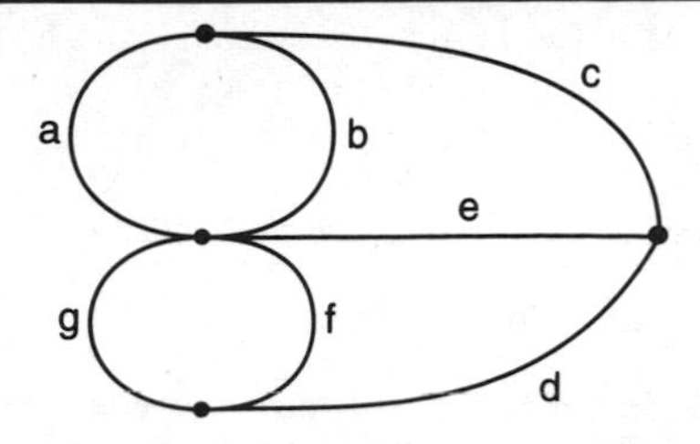

Fig. K2 *Königsberg bridge problem (graph)*

L

L Number 50 in the **Roman numeral** system.

l Abbreviation of **litre**.

laptop General name for a portable **computer**, usually of the same dimensions as a briefcase, and powered either by batteries where no mains electricity is available, or by mains electricity where it is available.

lateral velocity Component of the velocity of a celestial body that is at right angles to its **line-of-sight velocity**.

Latin square Group shown as a number of elements arranged in columns and rows, where every element occurs only once in each row and each column. *Example*:

a b c d
b a d c
d c a b
c d b a

latitude Imaginary line drawn parallel to the Equator around the Earth (a **small circle**). The Equator is 0° latitude, and the distance between the Equator and the poles is divided into 90° of latitude. Using latitude in combination with **longitude**, any position on the Earth's surface can be plotted. Alternative name: **parallel**.

lb Symbol for **pound**.

LCD Abbreviation for **lowest common denominator**.

LCM Abbreviation of **lowest common multiple**. Finding the LCM is one of the quickest ways of simplifying a range of **fractions**. *Example*: in the range of fractions $\frac{3}{4}$, $\frac{2}{5}$, $\frac{7}{10}$, and $\frac{13}{20}$ the LCM is 20. Therefore $\frac{3}{4}=\frac{15}{20}$; $\frac{2}{5}=\frac{8}{20}$; $\frac{7}{10}=\frac{14}{20}$. *See also* **lowest common denominator**.

leap year Year which is 366 **days** in length, as opposed to the normal 365 days. Leap years are those in which the number of the year can be divided by 4, e.g. 1988; 1992; 1996, and so on. Leap years are necessary because, although one **calendar year** is supposed to represent one complete passage of the Earth around the Sun, it in fact takes the Earth 365·259641 **solar days** to completely orbit the Sun. The solution arrived at is, in a sense, a kind of **rounding off**: the 0·25 of a day is carried on from one year to the next until the accumulated total equals one complete day. Since $0{\cdot}25 \times 4 = 1$, the complete day is reached every fourth year. If people did not follow this practice, the calendar year would gradually fall out of **synchronisation** with the visible movement of the Sun. Particular months which we now associate with the beginning and end of particular seasons would then gradually move out of synchronisation with those seasons. Try to calculate what the difference would be between the dates of the beginning of April in the year 2040 on the calendar incorporating leap years and on one which does not.

■ **length (unit of)** Any consistent unit employed to measure distance in one plane. The most commonly used units are: the **S.I. unit** of length, which is the **metre**; and the **British unit** of length, which is the **yard**. [2/5/h]

less than Approximate measure of difference between two numbers, shown by the symbol $<$. *Example*: $8 < 12$ means '8 is less than 12'. *Compare* **greater than.**

light pen Computer **input device**, used in association with special software, that enables an operator to write or draw on a VDU screen more or less as if using an ordinary pen.

limit In mathematics, value to which a **sequence** or **series** tends as more and more terms are included.

linear *1.* Involving only one **dimension**. *2.* Describing an **equation** of the first order. *3.* Describing a straight line.

■**linear equation** *1.* Algebraic equation of the first order, where all the values of x and y are of degree 1 (i.e. x and y are not squared or cubed etc. values) *Example*: $2(x-4y+8)=14$. [3/6/c] *2.* In **coordinate geometry**, the equation of a straight line. *See* **straight line, equation of**.

■**linear programming** Mathematical programming using linear inequalities. It can be used to find maximum and minimum values, for example in business when it is important to have a way of calculating maximum profits in a given situation, or calculating minimum costs or overheads where particular restraints are in operation. [5/10/c] *Example*: use the graph in Figure L1 to show the inequalities $x \geq 0$, $y \geq 5$ (the shaded area on the graph is the set required):

$$2x+3y \leq 30$$
$$y \geq 2x-4$$

■**line graph** Graph in which a single line is used to join the points plotted on the graph. The line can be either straight or curved, as in the graphs shown in Figure L2. [5/4/d]

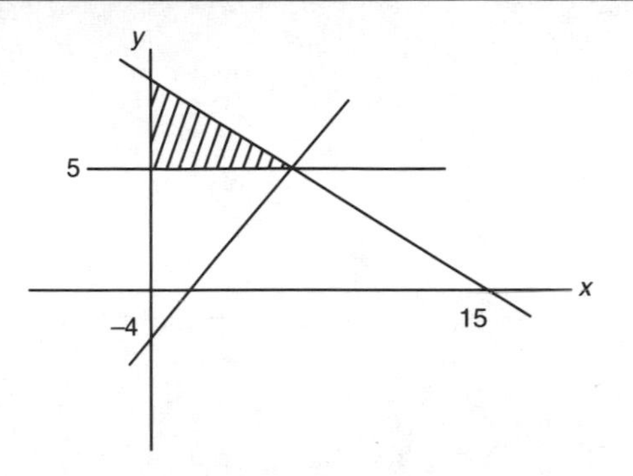

Fig. L1 *linear programming*

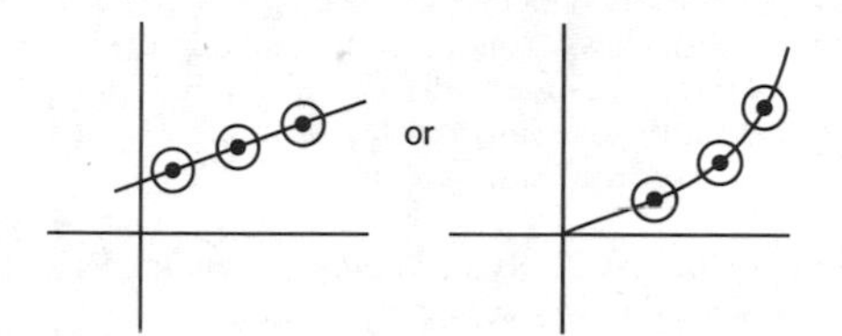

Fig. L2 *line graph*

■ **line of best fit** Line drawn on a **line graph** which denotes a trend in the graph, as shown in Figure L3, where the line is drawn through the points on the graph which all fall on the same line. A line of best fit is usually plotted when using experimental data, because this kind of data can vary widely. A line of best fit enables the experimenter to find the results which most fall along a similar path. [5/7/f]

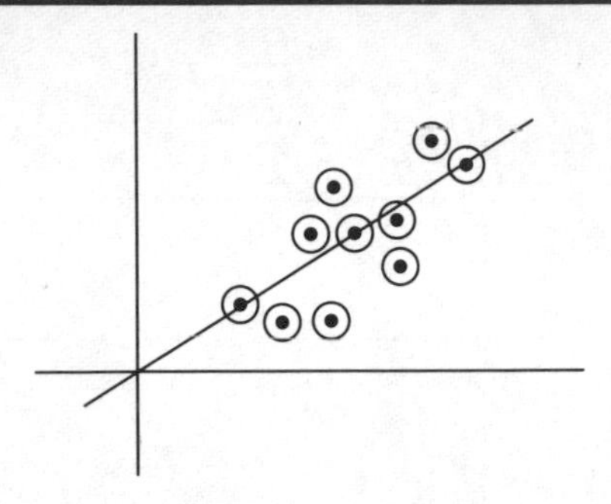

Fig. L3 *line of best fit*

line-of-sight velocity Velocity at which a celestial body moves towards or withdraws from (i.e. moves along the line of sight of) an observer. Because of the Earth's orbital motion, the velocity is usually standardized by being given as if the observer were on the Sun.

line printer Comparatively fast computer **output device** that prints out **data** one whole line at a time.

line segment Portion (or part) of a line.

■ **line symmetry** Exact reflection of a shape along a line. The line is often referred to as the **axis of symmetry**. [4/3/b]
Example: Figure L4 has two lines of symmetry.

Lissajous figures Closed plane curves formed by the combination of two or more simple periodic sinusoidal motions at right angles to each other, as shown in Figure L5,

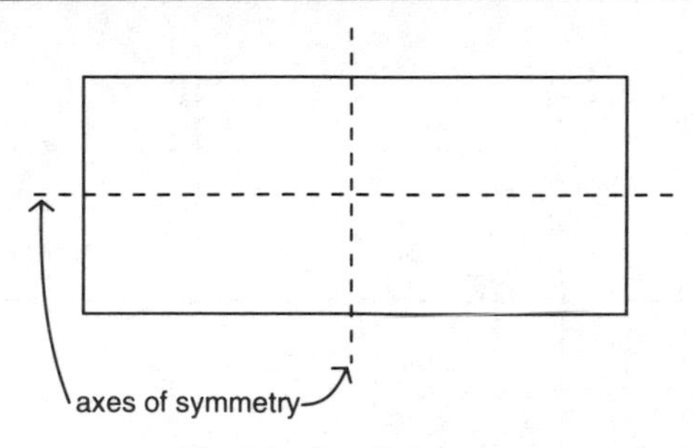

Fig. L4 *line symmetry*

where a and b are the respective amplitudes, ω and Ω are the angular frequencies, and α and β the initial phase angles. Lissajous figures are named after the French physicist Jules Lissajous (1822–80). The formulae are as follows:

$$x = a \sin(\omega t + \alpha)$$

and

$$y = b \sin(\Omega t + \beta).$$

Often the curves or figures produce closed loops which are determined by the ratio of the frequencies between the waves, as shown in Figure L5 where the curves for the ratios 1:2, 1:3, 2:3 and 3:4 are illustrated. This phenomenon is used with a cathode ray oscilloscope to compare one frequency with another. Alternative name: Lissajous' circle.

■**litre** (l) Unit of volume or capacity in the **S.I. unit** system. One litre = 1,000 ml = 1·76 pints. [2/5/i]

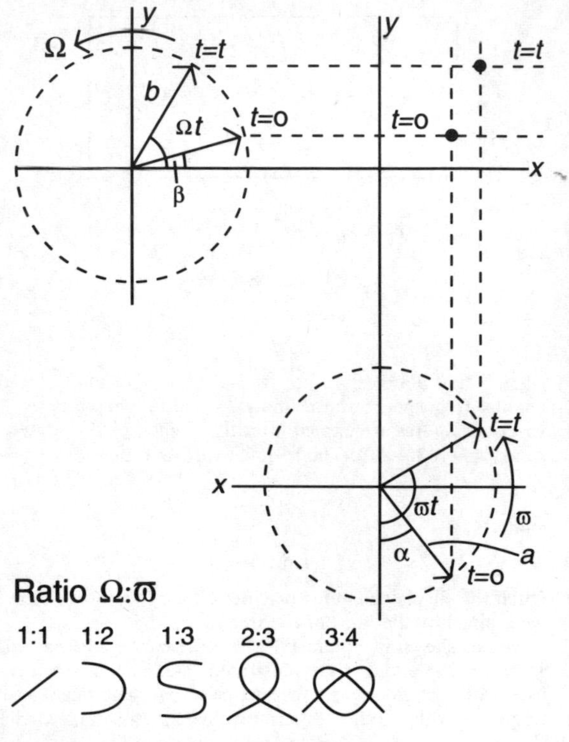

Fig. L5 *Lissajous figures*

ln Abbreviation for **Napierian** or **natural logarithm**.

loan Sum lent to a **borrower**, usually for a fixed period of time, and at a stated rate of **interest**. Loans are also often given on an agreed form of **security**. *See* **mortgage**.

loci Plural of **locus**.

■ **locus** (plural loci) In mathematics, path traced by a moving point; i.e. a line that can be drawn through adjacent positions of a point, each position of that point satisfying a particular set of conditions. [4/7/b] *Example*: the locus of a point that moves so that it is always the same distance from another fixed point is a **circle**.

log Abbreviation normally used for a **common logarithm**.

logarithm (log) **Power** to which a **base** must be raised to give an ordinary number. If y is a number and x is the base, $y = x^n$ where n is the logarithm of y to the base x. *Example*: the logarithm of 100 to the base 10 (the **common logarithm**) is 2, because 10^2 is 100. There are two basic types of logarithm: common logarithms are to the base 10, **Napierian logarithms** are to the base 2·71828 . . . Multiplication or division of ordinary numbers is achieved by the addition or subtraction of logarithms; the square of an ordinary number is found by multiplying the log of the number × 2; the cube of a number = log of the number × 3; the square root of a number = log of the number ÷ 2. Most **scientific calculators** will return common and natural logarithmic values. Logarithms were invented by the mathematician Napier as a way of trying to simplify calculations using large numbers. *Example*: until the widespread appearance of electronic calculators, when the only other option was using a pencil and a sheet of paper, it was much easier to express a number such as 10,000 as log 4 (10,000 to the base $10 = 10^4 = 4$), or to perform an operation such as 1182×610 using tables of logarithms than using, say, **long multiplication**. Expressed as a common logarithm operation

$$1182 \times 610 = 3{\cdot}0726 + 2{\cdot}7853 = 5{\cdot}8579.$$

(remember that multiplication using logarithms is achieved by adding the logs). To find the number which 5·8579 is the logarithm of, look up the **antilogarithm** table for ·8579, which gives 7·209. The **characteristic** is 5, therefore the answer is 720900; rounding up gives 721000. Use of logarithms equally makes the finding of a square root much easier. To find the square root of 1,278, the logarithm 3·1065 is found and then divided by 2 to give 1·5533, the antilogarithm of which is 35·75. To perform this last task using pencil and paper would be very time-consuming. After Napier's death, the use of **logarithm tables** as an aid to calculation spread widely. *See also* **natural logarithm**; **slide rule**.

logarithmic scale Scale of measurement based on the logarithmic function. For **common logarithms** (to the base 10), an increase of one unit represents a tenfold increase in the quantity measured. *Example*:

$$10^1 = 10$$

$$10^2 = 100$$

$$10^3 = 1000$$

$$10^4 = 10000$$

$$10^5 = 100000$$

See **slide rule**.

logarithm table (log table) Table giving the **logarithms** of the numbers from 10 to 99 inclusive. Note: the **mantissa** only of the logarithm is given on a log table. *Example*: in the table below, the **common logarithms** for 10–10·3, 11–11·3 and 12–12·3 are shown in mantissa form

	0	1	2	3
10	0000	0043	0086	0128
11	0414	0453	0492	0531
12	0792	0828	0864	0899

The horizontal range of log tables commonly covers all the values of a number from ·1 to ·9, thus the usual range for the number 11 would be from 11·00 to 11·9. The logs and values are only given to ·3 here for reasons of space. Log tables can be made for both Napierian and common logarithms: such tables usually state which type of logarithm is being used. Since the invention of the mechanical desk **calculator** and, more particularly, the electronic digital calculator, log tables have been less and less used. Until the invention of those devices, however, log tables were a frequently used aid to calculation in many fields of applied mathematics, from engineering to finance. Napier, the 16th century inventor of logarithms, was the first person to draw up log tables.

logic *1.* In mathematics, the use of methods from mathematics and formal logic to analyze the underlying principles on which mathematical systems are based. *2.* In electronic data-processing systems, the principles that define the interactions of data in the form of physical entities. *See also* **Boolean algebra**.

logic circuit Electronic switching circuit that gives an output only when specified input conditions are met. It is part of a

computer that performs a particular logical operation (e.g. 'and', 'or', 'not', etc.). Alternative names: logic element, **gate**. *Example*: Figure L6 shows the circuit symbols for five different logic circuits: AND gate, OR gate, NOT gate, NOR gate, and NAND gate respectively.

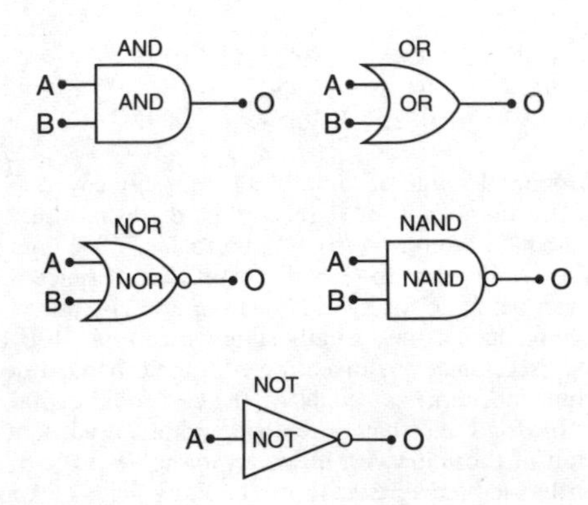

Fig. L6 *logic circuits*

Truth tables can be drawn up for these, as follows:

INPUTS		*OUTPUTS*				
A	B	AND	OR	NOT	NOR	NAND
0	0	0	0		1	1
0	1	0	1	1	0	1
1	0	0	1	0	0	1
1	1	1	1		0	0

log table Abbreviation for **logarithm table**.

long division Method of calculation where the operation is carried out line-by-line, digit-by-digit, carefully following **place-value notation**. *Example*: to divide the **denary** terms 36,784 ÷ 129, the divisor is divided into the dividend as many times as it can be equally, subtracting the multiple of quotient and divisor from the dividend at each complete division:

```
        285
129 | 36784
      258
      1098
      1032
       664
       645
        19
```

The quotient, therefore, is $285\frac{19}{129} = 285{\cdot}14728$.

longitude Imaginary line that passes halfway around the Earth through both poles. The angular distance around the globe is 360°, which is measured as 180° east of Greenwich (designated 0°) and 180° west of Greenwich. Using longitude in combination with **latitude**, any position on the Earth's surface can be plotted. Alternative name: **meridian**.

long multiplication Method of calculation where the operation is performed line-by-line, digit-by-digit, carefully following **place-value notation**. *Example*: to multiply the **denary** terms 812 and 118 (812 × 118), first place the multiplicand directly above the multiplier

```
  812
× 118
```

Now, multiply 812 by 8 (the units number) = 6496; place this result directly below the initial expression

```
  812
× 118
─────
 6496
```

Next, multiply 812 by the tens value (1 in this case), and then by the hundreds value (again, 1). Place these results directly below the previous expression, taking care to place the tens answer in the tens column, and the hundreds answer in the hundreds column, to give

```
  812
× 118
─────
 6496
  812
 812
```

Now carefully add the three lines which are the result of the multiplication, ensuring that units are added to units, tens to tens, and hundreds to hundreds

```
  812
× 118
─────
 6496
  812
 812
─────
95816
```

The answer is 95,816.

Lorentz transformation Set of equations for correlating space and time coordinates in two frames of reference.

■ **lower bound** Minimum extreme or value of a particular quantity, e.g. the lower bound of the trigonometrical **sine** function is −1. [2/9/b]

■ **lowest common denominator** (LCD) Number that is the **lowest common multiple** of all the denominators of a set of **fractions**. It is often necessary to find the LCM in order to add or subtract fractions. [2/5/d] *Example*: the lowest common denominator of the fractions $\frac{1}{3}$, $\frac{1}{4}$, and $\frac{1}{2}$ is 12. Therefore $\frac{1}{4}=\frac{3}{12}$; $\frac{1}{3}=\frac{4}{12}$ and $\frac{1}{2}=\frac{6}{12}$:

$$\begin{aligned}\tfrac{1}{3}-\tfrac{1}{4}+\tfrac{1}{2}&=\tfrac{4}{12}-\tfrac{3}{12}+\tfrac{6}{12}\\&=\frac{4-3+6}{12}\\&=\tfrac{7}{12}\end{aligned}$$

See also **reduction**.

lowest common multiple (LCM) Smallest number that all the members of a group of numbers will divide into. E.g. the lowest common multiple of 2, 4 and 5 is 20.

■ **lowest terms (of a fraction)** Terms below which a fraction cannot be simplified because it is not possible to find a number that will divide into both the numerator and the denominator, e.g. $\frac{3}{8}$; $\frac{16}{21}$; $\frac{221}{363}$. [2/5/d]

M

M *1.* Abbreviation for **mile.** *2.* Symbol for **mega-**, $\times 10^6$. *3.* Number 1,000 in **Roman numeral** system.

m *1.* Abbreviation for **mean.** *2.* Abbreviation for **metre.** *3.* Symbol for **milli-**, $\times 10^{-3}$.

m Symbol for **mass.**

machine code Code in which instructions are given to a **computer.** Many computer languages (used for **programs**) have to be translated into machine code before they can be 'understood' by a computer.

magic square Square array of numbers whose columns, rows and diagonals add to the same total, as shown below.

16	10	7	1
3	5	12	14
2	8	9	15
13	11	6	4

magnetic core Computer storage device consisting of a ferromagnetic ring wound with wires; a current flowing in the wires polarizes the core, which can therefore adopt one of two states (positive or negative).

magnetic disk Device for direct-access storage and retrieval of data, used in computers and similar systems. It consists of a rotatable flexible or rigid plastic disc (i.e. a floppy or hard disk) coated on one or both surfaces with magnetic material, such as iron oxide. Data is stored or retrieved through one or more **read/write heads**. Alternative name: magnetic disc.

magnetic drum Computer storage device consisting of a rotatable drum coated with magnetic material, such as iron oxide. Data is stored or retrieved through one or more **read/write heads**.

magnetic tape Medium for the storage of electronic signals by magnetizing particles (of e.g. iron oxide) in a coating on plastic tape. It is used in (audio) tape recorders, video recorders and computers.

magnitude In mathematics, the numerical value of a **vector**; if a vector is represented by a line segment, its magnitude is the length of the line.

mainframe Increasingly outdated term for physically large **computers** with complex **architecture**, large memory capacity and fast processing speeds.

major arc **Arc** on a circle equal to more than half the length of its **circumference**. *Compare* **minor arc**. *See* **radian**.

major axis Longer of two **axes of symmetry** of a geometric figure (e.g. of an **ellipse**). The other axis is the minor axis.

major sector Sector of a **circle** greater in area than a semicircle. *See* **radian**.

mantissa Fractional part of a **logarithm** (the other part is the **characteristic**); e.g. in the logarithm 2·3010, ·3010 is the mantissa, and 2 is the characteristic.

■**mapping** Transformation of mathematical information from one form to another. A correspondence can be established between two sets of elements if to each element in one set there is a unique corresponding element in the other. [3/5/b]
Example: x can be mapped onto $x+2$ for values of x between 0 and 4 as follows:

$$x=0: x+2=0+2=2$$

$$x=1: x+2=1+2=3$$

$$x=2: x+2=2+2=4$$

$$x=3: x+2=3+2=5$$

$$x=4: x+2=4+2=6$$

This can be shown diagrammatically as:

x	$x+2$
0	2
1	3
2	4
3	5

mass (m) Quantity of matter in an object, and a measure of the extent to which it resists **acceleration** if acted on by a force. The **SI unit** of mass is the **kilogram** (kg). *See also* **mechanics**.

mathematics Branch of science concerned with the study of numbers, quantities and space; broadly speaking, it can be divided into the different but related areas of **arithmetic**, **geometry**, **trigonometry** and **algebra**. The origins of mathematics cannot be precisely identified, but evidence from all cultures, whether agricultural, nomadic, settled, hunter-gatherer or otherwise suggests that human beings have since the earliest times used ways of describing quantities of things, and have also invented ways of performing the operations of addition, subtraction, multiplication and division. Although the growth of Western mathematics has been dominated by the mathematical discoveries and inventions of ancient peoples such as the Babylonians, the Egyptians and the Greeks, other cultures, equally ancient, also developed their own ways of performing complex calculations. The civilizations of ancient China and India both possessed highly evolved systems for, for example, calculating the distances and movements of the stars and planets, as well as very complex systems of notation for writing down numbers and performing calculations. The same is true of the highly-developed ancient cultures of South America, such as the Mayan culture. It is difficult to identify particular human activities which spurred on the development of mathematical ideas. It has been suggested, for example, that counting domesticated animals in herds may have been important in this respect. Creating buildings; observing the stars and planets; dividing up land for territorial purposes and creating calendars are all, however, certainly tasks which demand carefully thought-out systems of understanding and using numbers. Modern mathematics is generally dominated by the Western tradition, and the invention of **computers** has given mathematicians calculating devices which are much more powerful than any which existed before. These have enabled people to plan and execute tasks which have tested the accuracy of modern mathematics very rigorously, for example the launching and planned re-entry into the Earth's atmosphere of spacecraft. Partly also because of the availability of computers, mathematicians now study extremely complex continuously-developing phenomena such as the Earth's weather systems, trying to discover mathematical ways of describing how the

Earth's air and water systems interact with each other. One aspect of such studies has been the development of interest in what is termed 'chaos theory', a series of attempts to discover mathematical principles underlying apparently totally unrelated events, for example how the weather and ground conditions in one part of the world can powerfully affect those in another, perhaps thousands of miles away.

■**matrix** In mathematics, square or rectangular array of elements, e.g. numbers. Matrices are often used when a large amount of numerical data has to be manipulated, or an array of numbers has been used. [4/10/d] *Example*: if a collection of compact discs were placed in 3 racks under the headings of classical, jazz, pop and rock, they could be written as:

racks	*classical*	*jazz*	*pop*	*rock*
A	6	4	2	1
B	3	3	8	2
C	2	1	5	6

After a time, the headings could be forgotten, and the information could be written as:

$$\begin{pmatrix} 6 & 4 & 2 & 1 \\ 3 & 3 & 8 & 2 \\ 2 & 1 & 5 & 6 \end{pmatrix}$$

i.e., an array of numbers.

matrix notation Arrangement of numbers (known as elements) placed in rows and columns. A matrix is always drawn in closed brackets. The order of a matrix can be specified. If it has 5 rows and 4 columns it is of order 5×4.

The number of rows is always placed first. There are various types of matrix:

(a) *row matrix*: a matrix simply having a row; no columns, e.g. (5 4).

(b) *column matrix*: a matrix having one column and no rows, e.g. $\begin{pmatrix} 5 \\ 4 \end{pmatrix}$

(c) *square matrix*: a matrix having the same number of rows as columns, e.g. $\begin{pmatrix} 5 & 3 \\ 4 & 2 \end{pmatrix}$

(d) *null matrix*: a matrix with all the elements equal to zero, e.g. $\begin{pmatrix} 0 & 0 \\ 0 & 0 \end{pmatrix}$

(e) *unit matrix*: denoted by $\begin{pmatrix} 1 & 0 \\ 0 & 1 \end{pmatrix}$

usually denoted by the symbol I. For addition of matrices, see **addition of matrices**. *Subtraction of matrices* is achieved in a way similar to addition:

$$\begin{pmatrix} a & b \\ c & d \end{pmatrix} - \begin{pmatrix} e & f \\ g & h \end{pmatrix} = \begin{pmatrix} a-e & b-f \\ c-g & d-h \end{pmatrix}$$

or, to use numbers in place of letters:

$$\begin{pmatrix} 1 & 2 \\ 3 & 4 \end{pmatrix} - \begin{pmatrix} 5 & 6 \\ 7 & 8 \end{pmatrix} = \begin{pmatrix} -4 & -4 \\ -4 & -4 \end{pmatrix}$$

Multiplication of matrices can be of two kinds: scalar multiplication, and matrix multiplication. Scalar multiplication, where a matrix is simply multiplied by a number is achieved as follows:

$$3\begin{pmatrix}1 & 2\\ 3 & 4\end{pmatrix} - \begin{pmatrix}3\times 1 & 3\times 2\\ 3\times 3 & 3\times 4\end{pmatrix} = \begin{pmatrix}3 & 6\\ 9 & 12\end{pmatrix}$$

Matrix multiplication can only take place if the number of rows and columns in each matrix is the same. If they are, then a row is multiplied by a column:

$$\begin{pmatrix}2 & 3\\ 4 & 5\end{pmatrix} \times \begin{pmatrix}5 & 2\\ 3 & 6\end{pmatrix} = \begin{pmatrix}2\times 5+3\times 3 & 2\times 2+3\times 6\\ 4\times 5+5\times 3 & 4\times 2+5\times 6\end{pmatrix}$$

$$= \begin{pmatrix}10+9 & 4+18\\ 20+15 & 8+30\end{pmatrix}$$

$$= \begin{pmatrix}19 & 22\\ 35 & 38\end{pmatrix}$$

max Abbreviation for **maximum**.

maximum (max) *1.* In **coordinate geometry**, the point on a curve that represents the largest value of a **function**. *Compare* **minimum**. *2.* Largest value in a group of numbers.

■**mean** (m) Average of a group of numbers, for example in the range of numbers 6, 22, 14, 8, 12, 22, the **arithmetic mean** is equal to the sum of all the numbers divided by the number of numbers, i.e.

$$\frac{6+22+14+8+12+22}{6} = \frac{84}{6}$$
$$= 14$$

[5/4/f] [5/7/c] *See also* **geometric mean**.

mean deviation Sum of all deviations from the **arithmetic mean** of a group of numbers, divided by the quantity of numbers in the group.

measure System of regular **units**. The name given to the system and/or units is generally speaking not important for measurement in itself, although it is important when giving measurements to be clear what units are being used. The essential element is that the units employed are regular and consistent one with the other. *Example*: a line measured in **metric** units may equal 45 cm; the same line will equal 1·47 feet, or 1′ 5·71″ on the **f.p.s. system**.

measures of dispersion Amount of scatter in a set of data; usually measured by the **standard deviation**. See also **scatter diagram**.

mechanics Study of the interaction between matter and the forces acting on it. Its three divisions are: **kinematics** (concerned with acceleration, velocity, etc.); **dynamics** (concerned with objects in motion); and statics (concerned with forces that do not produce motion).

■ **median** Value of the middle number (or the average of the two middle numbers if there is no single middle number) of a group of numbers arranged in ascending order. [5/7/c] *See also* **mode**.

mega- (M) Prefix in the **metric** system denoting one million ($\times 10^6$).

member One of a collection of things which share one or more characteristics. In mathematics, the word usually refers to the parts or elements of a **set**.

memory Part of a computer that stores data and instructions (programs), usually referring to the immediate access store. *See also* **random access memory** (RAM); **read-only memory** (ROM).

mensuration Science of measurement of geometric figures, for example the calculation of **area** and **volume**.

Mercator's projection Cylindrical projection of the Earth's surface onto a flat sheet, the most familiar world map. It has long been used by navigators because it shows correct distance and bearing, although it is in some ways misleading because it greatly exaggerates the area of land masses near the poles. It was named after the Flemish cartographer Gerardus Mercator (1512–94). *See also* Appendix III.

meridian Line of **longitude**, half of a **great circle** that joins the poles.

■**metre** (m) **S.I. unit** of length; 100 cm = 1 m = 3·28 **feet**. [2/5/i]

metric prefix Any of various numerical prefixes used in the **metric system**.

Prefix	*Symbol*	*Multiple*	*Prefix*	*Symbol*	*Multiple*
atto-	a	$\times 10^{-18}$	deca-	da	$\times 10$
femto-	f	$\times 10^{-15}$	hecto-	h	$\times 10^{2}$
pico-	p	$\times 10^{-12}$	kilo-	k	$\times 10^{3}$
nano-	n	$\times 10^{-9}$	mega-	M	$\times 10^{6}$
micro-	μ	$\times 10^{-6}$	giga-	G	$\times 10^{9}$
milli-	m	$\times 10^{-3}$	tera-	T	$\times 10^{12}$
centi-	c	$\times 10^{-2}$	peta-	P	$\times 10^{15}$
deci-	d	$\times 10^{-1}$	exa-	E	$\times 10^{18}$

metric system Decimal-based system of units.

micro- Prefix in the **metric system** signifying $\times 10^{-6}$.

microprocessor *See* **computer**.

■ **mid-ordinate rule** Rule for finding the area under a curve, bounded by a horizontal below the curve. A vertical strip is drawn between the curve and the horizontal below it. The vertical line midway between the extreme vertical edges of the strip is called the mid-ordinate. If the area between the curve and the horizontal axis is divided up into strips of equal width, and the height of the mid-ordinate for each strip is found, the area under the curve is equal to the width of the strip multiplied by the sum of all the mid-ordinates. [3/10/c] *Example*: in Figure M1, the mid-ordinates are m_1, m_2, m_3 and m_4; the distance between each strip is represented by b. The area under the curve is:

$$bm_1 + bm_2 + bm_3 + bm_4 = b(m_1 + m_2 + m_3 + m_4)$$

■ **mile** (M) **British unit** of length. The statute mile is the Standard British unit of length: 1 mile = 8 furlongs = 1,760 yards = 1·609 km. The geographical mile is equal to one **minute** of arc measured at the equator of the Earth, equal to 6,087·2 feet = 1·85 km. The nautical mile is the unit of length used for navigation. It is the length of a minute of arc on a **great circle** drawn round the Earth, approximated to 6,080 feet due to the flattening of the poles. [2/5/h] *See also* **knot**.

milli- Prefix on the **metric system** signifying $\times 10^{-3}$.

millilitre (ml) **SI unit** of liquid measure equal to one thousandth part of a **litre** = 0·0351 **fluid ounces**.

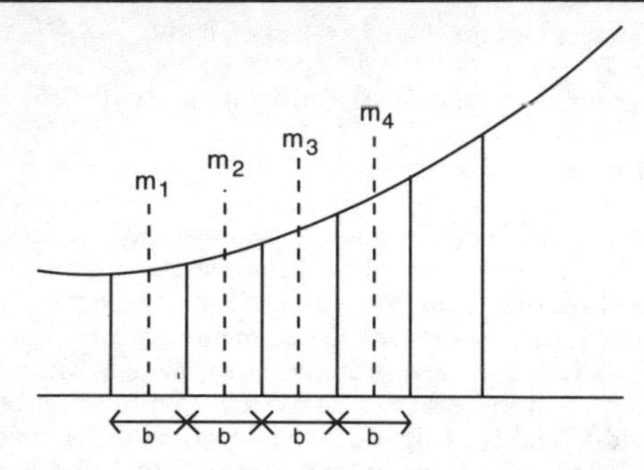

Fig. M1 *mid-ordinate rule*

■**millimetre** (mm) **S.I. unit** of length, equal to one thousandth part of a metre. [2/5/i]

million Number equal to 1,000,000 (10^6).

min *1.* Abbreviation for **minimum**. *2.* Abbreviation for **minute**.

minimum (min) *1.* In **coordinate geometry**, a point on a curve where its slope changes from negative to positive, as shown in Figure M2 (a tangent to the curve at that point has a slope of zero, and therefore the **function** = 0). *Compare* **maximum**. *2.* The smallest value in a group of numbers.

■**minor arc** Arc on a circle that is less than half the **circumference**, as shown in Figure M3. [4/9/d] *See* **major arc**; **radian**.

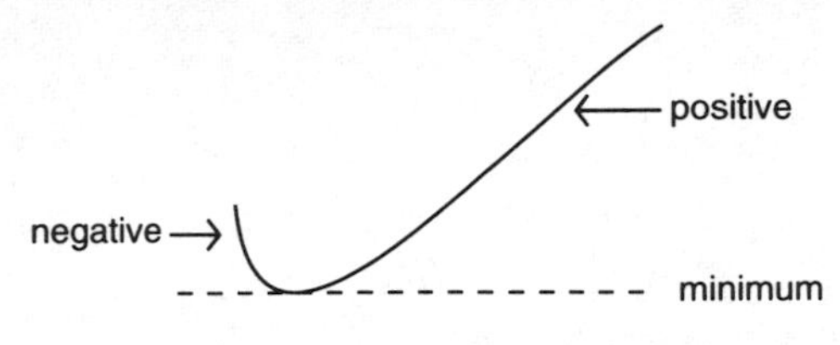

Fig. M2 *minimum* (*sense 1.*)

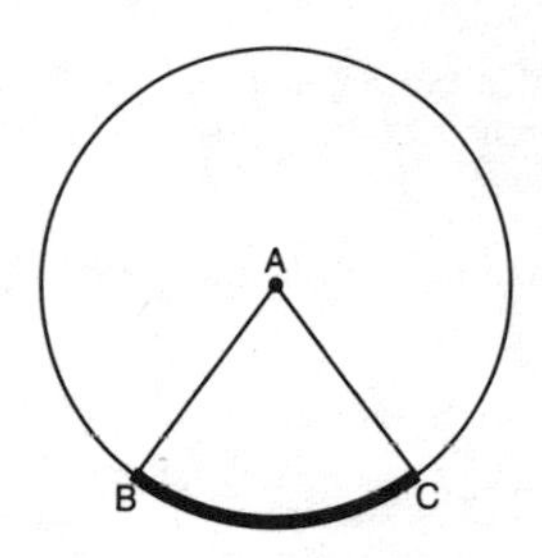

Fig. M3 *minor arc*

minor axis Shorter of the two axes of a symmetrical geometrical figure (e.g. of an **ellipse**). The other axis is the **major axis**.

■**minor sector** Sector of a **circle** that is less than a semicircle, as shown in Figure M4. *See* **radian**. [4/9/d]

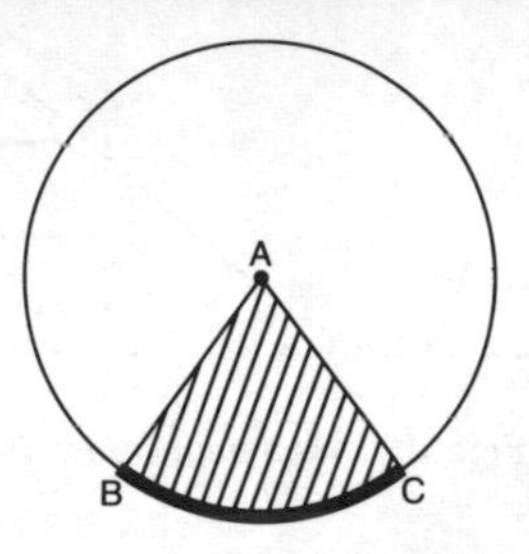

Fig. M4 *minor sector*

minute (min) *1.* Unit of angular measure equal to 1/60 of a degree and made up of 60 seconds. *2.* Unit measure of time; 60 minutes = 1 **hour**.

mixed number Sum of a whole number (integer) and a fraction, e.g. $3\frac{2}{3}$; $12\frac{15}{16}$.

ml Abbreviation for **millilitre**.

mm Abbreviation of **millimetre**.

Möbius strip Continuous circular band that has only one flat surface and one edge, formed by rotating one end of a rectangle (e.g. a strip of paper) through 180° before joining the ends. It was named after the German mathematician August Möbius (1790–1868). *See also* Appendix III.

mod Abbreviation of **modulus**.

■**mode** Number in a group of numbers that occurs most frequently. [5/7/c] *See also* **median**.

modem Acronym of modulator/demodulator, a device for transmitting computer **data** over long distances (e.g. by telephone line).

modular arithmetic Arithmetic performed using numbers determined in a certain base system. If we use base 5 this could be abbreviated to modulo 5. The modulo n set uses only 0, 1, 2, . . . $n-1$. All mathematical operations are the same except when the number chosen is greater than $n-1$. Then it is divided by n and the remainder is used as the required answer. *Example*: 24 in modulus 5 would become 24/5 =5 remainder 4. Therefore 24 in modulo 5 is 4.

modulator/demodulator *See* **modem**.

modulus *1.* Value of a **real number** irrespective of its sign (positive or negative); abbreviation **mod**. Alternative name: **absolute value**. *2.* The value $\sqrt{(x^2+y^2)}$ of a **complex number** $x+iy$. On an **Argand diagram**, it is the distance from the origin to the point representing the complex number.

mol Abbreviation for **mole**.

mole (mol) **SI unit** of amount of substance.

moment of force Product of the perpendicular distance of an **axis** from the line of action of the force and the **component** of the force, in the plane perpendicular to the axis. It has a turning effect (**torque**).

moment of inertia Sum of the products of the mass of each particle of a body about an **axis** and the square of its perpendicular distance from the axis (its **radius of gyration**). Symbol I.

momentum Product of the mass m and velocity v of a moving object (symbol p); i.e. $p = mv$. It is a **vector** quantity directed through the centre of mass of the object in the direction of motion. When objects collide, the total momentum before impact is the same as the total momentum after impact.

month Time period of between 28 and 31 **days**, based on the period of time it takes the Moon to complete one orbit of the Earth (27·32 days). Although the names for the different months vary from country to country around the world, there are always 12 months in a **calendar year**.

mortgage Loan secured on the title deeds of a property. A purchaser borrows the sum required to buy a property; in return the lender (most often a building society) requires to be given the title deeds, which will be held until the loan is fully repaid. *See* **security**.

mph Abbreviation for **miles** per hour.

m/s Abbreviation for **metres** per second.

multiple Number or algebraic expression that is an exact number of times bigger than another number or expression. *Example*: (a) 24 is a multiple of 6; (b) $12x + 8y$ is a multiple of $3x + 2y$.

■ **multiplication** Increasing a number by another number, each of which is termed a **factor**. The result is the **product** of two factors. [2/3/c,d] The process is denoted by (a) a × sign; (b) a dot (.); or (c) closing terms up to each other. The order in which the two factors are multiplied does not matter:

$$\text{(a)}\quad a \times b = c$$

$$\text{(b)}\quad a \,.\, b = c$$

$$\text{(c)}\quad ab = c$$

multiplication law of probability Multiplication of the individual probabilities of two or more independent events to find the probability that all these independent events will take place.

multiplication of fractions If fractions are multiplied together the numerators are multiplied together and so are the denominators. *Example*:

$$\tfrac{3}{4} \times \tfrac{2}{3} = \frac{3 \times 2}{4 \times 3} = \tfrac{6}{12} = \tfrac{1}{2}$$

■ **mutually exclusive** Describes events that could not occur together, i.e. the occurrence of one renders the simultaneous occurrence of the other impossible. If the separate probabilities are found then the probability of one of the events happening is the sum of the individual probabilities. [5/7/i] *Example*: What is the probability of cutting the ace of spades, king of spades or the jack of spades in one cut? The probabilities can be calculated as follows:

$$\text{P (ace)} = P_1 = \tfrac{1}{52}$$

$$\text{P (king)} = P_2 = \tfrac{1}{52}$$

$$\text{P (jack)} = P_3 = \tfrac{1}{52}$$

Therefore, the probability is:

$$P_1 + P_2 + P_3 = \tfrac{1}{52} + \tfrac{1}{52} + \tfrac{1}{52} = \tfrac{3}{52}$$

N

N *1.* Symbol for **newton**. *2.* Abbreviation for **compass** direction North.

n Symbol for **nano-**, $\times 10^{-9}$.

nadir In astronomy, the point on the **celestial sphere** that is directly below an observer's feet. *See also* **zenith.**

nano- (n) Prefix in the **metric system** signifying $\times 10^{-9}$.

Napierian logarithm (ln) **Logarithm** to the base e (e = 2·71828 . . .), named after the Scottish mathematician John Napier (1550–1617). Most **scientific calculators** will return Napierian logarithmic values; the relevant key is usually marked 'ln'. Alternative name: **natural logarithm**. *Compare* **common logarithm**; *see also* Appendix III.

natural logarithm (ln) Alternative name for **Napierian logarithm.**

natural number One of the set of ordinary counting numbers (*e.g.* 1, 2, 3, 4, etc.).

nautical mile *See* **mile.**

necessary condition Important part of a mathematical rule which must hold if the rule is being applied. *Example*: it is a necessary condition of a **cube** that all six **faces** are equal in **area**.

■ **negative number** Number less than zero. [2/5/j]

■**net** *1.* Figure created when a solid is opened out so that all the faces of it lie in one plane. [4/5/a] *Example*: the net of a **cube** would look as shown in Figure N1. *2.* Amount after agreed deductions are made. *Example*: a person's **gross** salary is £15,000 **per annum**. Annual tax of £3,500 and National Insurance contributions of £1,250 are deducted to give a net salary of 15000 − (3500 + 1250) = £10,250.

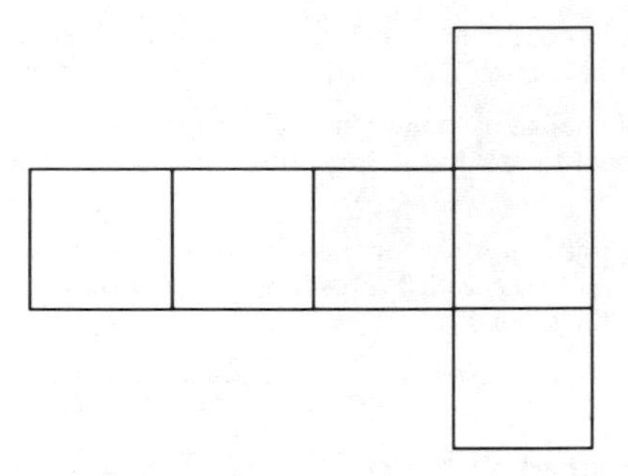

Fig. N1 *net* (*sense 1.*)

network In computing, a system of interconnected terminals on-line to one or more computers.

■**newton** (N) **S.I. unit** of force. It is the force required to produce an **acceleration** of 1 m s^{-2} (1 metre per second^2) on a **mass** of 1 kilogram. [2/5/i]

Newtonian mechanics System of mechanics that relies on **Newton's laws of motion** and is applicable to objects moving at speeds relative to the observer that are small compared to the speed of light. Objects moving near to the speed of light require an approach based on relativistic mechanics, in which the mass of the object changes with its speed (see **relativity**).

Newton's laws of motion Three laws of motion on which **Newtonian mechanics** is based:

1. an object continues in a state of rest or uniform motion in a straight line unless it is acted upon by external forces;

2. rate of change of momentum of a moving object is proportional to and in the same direction as the force acting on it;

3. if one object exerts a force on another, there is an equal and opposite force, called a reaction, exerted on the first object by the second.

Alternative name: Newton's laws of force. *See* Appendix III.

Newton's method Method of obtaining rough approximations to the values of the roots of an equation. *Example*: let $x=a$ be the assumed approximate root of an equation $f(x)=0$; let accurate roots be $x=a+d$ where d is very small. Therefore $f(a+d)=0$. Use of **differentiation** gives:

$$f'(a)=\lim_{d\to 0}\left\{\frac{f(a+d)-f(a)}{d}\right\}$$

$$\approx\frac{f(a+d)-f(a)}{d}$$

if d is small.

$$d \approx \frac{f(a+d)-f(a)}{f'(a)}$$

Therefore as $f(a+d)=0$

$$d \approx \frac{-f(a)}{f'(a)}$$

Therefore a closer approximation to the root would be:

$$x=a-\frac{f(a)}{f'(a)}$$

node Point on a curve that has two real and discrete tangents, often called the double point. *Example*: the folium curve of Descartes has the equation $x^3+y^3=3axy$ where the point (0,0) is a node, as seen in Figure N2, where the origin is the case for the node or double point. A nodal point is part of a **tessellation** where there is a common vertex to three or more polygons.

nona- Prefix signifying nine, as in **nonagon**, or ninety, as in nonagenarian (a person reaching ninety years of age).

nonagon Nine-sided plane figure or **polygon**. Each of the **internal angles** of a regular nonagon equals 140°.

normal In mathematics, a plane or line that is perpendicular to another. At any point on a curve, the normal is perpendicular to the tangent to the curve at that point.

■**normal distribution Distribution** that fits a normal frequency curve. [5/10/b]

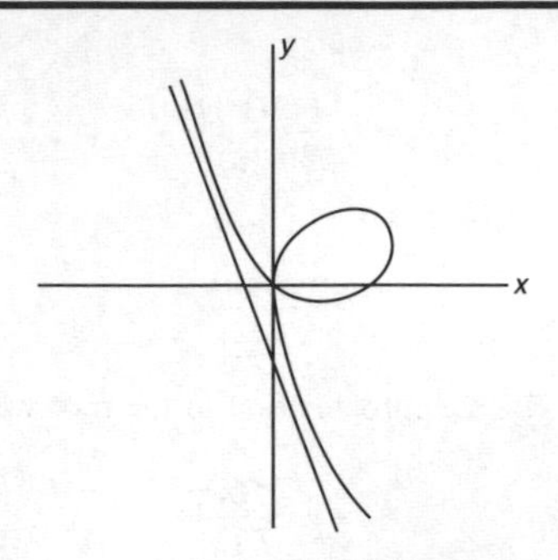

Fig. N2 *node* (*folium curve of Descartes*)

null hypothesis Proposition, used with contingency tables in statistical analysis, that the two analyses of data represented by columns and rows are statistically independent from each other. Often this method becomes unworkable as more and more data builds up.

null matrix Matrix with all the elements equal to zero:

$$\begin{pmatrix} 0 & 0 \\ 0 & 0 \end{pmatrix}$$

null set (symbol $\varnothing$) Set containing no elements, a subset of every set. *See* **empty set**; **proper subset**.

numeral In mathematics, symbol that represents a number. *Example*: (a) Arabic numerals 1, 2, 3, 4, 5, etc.; (b) **Roman numerals** I, II, III, IV, V, etc.

numerator Top part of a **fraction** (the lower part is the **denominator**).

O

■**oblong** Imprecise alternative name for **rectangle**. [4/2/a]

■**obtuse angle** Describing an angle that is more than 90° but less than 180°. [4/4/c]

OCR Abbreviation of **optical character recognition**.

OCT Abbreviation for **octal**.

octa- Prefix signifying eight, as in **octagon**, or eighty, as in octogenarian, a person reaching eighty years of age.

■**octagon** Eight-sided plane figure or **polygon**. All the interior angles of a regular octagon equal 135°. [4/2/a]

octal (OCT) Describing a number system to the **base** 8, widely used in computing. Most **scientific calculators** allow the user to perform operations in octal, and to translate from other number systems to octal and vice-versa. *Example*: 365·87 in **decimal** = 556 in octal.

octahedron Polyhedron with eight plane faces. A regular octahedron resembles two square-based pyramids base to base and has eight **equilateral** triangles as sides.

odd number Describing an integer which when divided by 2 gives a remainder of 1, for example, 3, 5, 7, 11, 239, 1181.

■**ogive** Curve obtained in a **cumulative frequency** distribution. [5/8/b]

on-line Describing part of a **computer** (e.g. an **input device**) that is linked directly to and under the control of the **central processor**.

operation Process carried out using set mathematical rules, e.g. addition or multiplication. **Associative**, **distributive** and **commutative** rules apply. *Example*: for a set of rational numbers x, y, z

Associative rule:

$$x+(y+z)=(x+y)+z$$

$$x(yz)=(xy)z$$

Distributive rule:

$$x(y+z)=(xy)+(xz)$$

$$(x+y)z=(xz)+(yz)$$

Commutative rule: $x+y=y+x$;

$$xy=yx$$

operator In mathematics, symbol or term that represents a mathematical operation to be carried out on a particular operand. *Example*: in the expression $a(x+2)$, a is the operator which implies multiplication.

opposite angle Angle in a triangle immediately opposite a given side. *Example*: in a right-angled triangle, the right angle is the opposite angle to the **hypotenuse**.

optical character reader Computer **input device** that 'reads' printed or written **alphanumeric** characters and feeds the information into a computer system. *See also* **character recognition**.

optical character recognition (OCR) Technique that uses an **optical character reader**.

order Integer used when describing **equations** or **powers**. *See* **order of equations**.

ordered pair Set of elements written in the sequence in which they are to be used. *Example*: the Cartesian coordinates (x, y).

■ **order of accuracy** Closeness of an approximate answer to the exact answer. It is often referred to as correct to n decimal places, where n is the order of accuracy. [2/10/a] [2/6/g]

order of a matrix Number of rows in a **matrix** multiplied by the number of columns.

order of equations Number of the highest power represented in an **equation**. There are first and second order differential equations. *See* **linear equation**.

order of symmetry In geometry, the number of times a figure is rotated until it coincides with its original position. *Example*: for a nth sided figure this would equal $360/n$.

ordinal number Number that indicates the rank of a quantity; e.g. 1st, 2nd, 3rd, etc. (as opposed to ordinary counting or **cardinal numbers**: 1, 2, 3, etc.).

ordinate In **coordinate geometry**, the y coordinate of a point (distance to the x-axis). The other (x) coordinate is the **abscissa**.

origin In **coordinate geometry**, the point where the x- and y-

axes cross (and from which **Cartesian coordinates** are measured).

orthocentre Point of intersection of the **altitudes** of a triangle.

orthogonal Line at right angles to another.

■**ounce** (oz) Unit measure of weight in the **f.p.s. system**, equal to one sixteenth of a **pound** = 28·35 **grams**. [2/5/h]

outcome Result or answer from a mathematical operation or sequence of operations.

output device Part of a **computer** that presents **data** in a form that can be used by a human operator; e.g. a **printer**, **visual display unit** (VDU), chart plotter, etc. A machine that writes data onto a portable magnetic medium (e.g. magnetic disk or tape) may also be considered to be an output device.

oval Geometric shape similar to a flattened circle; eggs are approximately oval in shape.

overdraft Short-term borrowing facility on a **current account** within limits agreed between a bank and one of its account holders; overdrafts attract **interest** charges.

oz Abbreviation for **ounce**.

P

P *1.* Abbreviation for **principal**. *2.* Symbol for **peta-**, $\times 10^{15}$.

p Symbol for **pico-**, $\times 10^{-12}$.

p.a. Abbreviation for **per annum**.

paper tape *See* **punched tape**.

parabola Curve traced by points that are separated by equal distances from a given fixed point (**focus**) and a straight line (**directrix**). *See* **conic section**. A parabola thus has an **eccentricity** of 1. Most **projectiles** follow parabolic paths; a parabolic **mirror** produces a parallel beam of light from a small light source placed at its focus; in a similar way, a parabolic dish aerial transmits or receives a parallel beam of microwaves.

parabolic Shaped like a **parabola**.

parallax In astronomy, the angle between the straight lines joining two vicwpoints to a heavenly body.

parallel Alternative name for **latitude**.

parallelepiped Solid figure that has six **faces**, i.e. all **parallelograms**, with opposite pairs of faces identical and parallel.

■ **parallel lines** Lines which have no intersection however far they are extended, and always have a constant distance between them. [4/5/b]

■**parallelogram** Four-sided **plane** figure whose opposite sides are equal and parallel. Its area equals the **product** of its length and its height (altitude), i.e. $A = l \times h$. [1/5/b]

parallelogram of forces Method of finding a **resultant** of two forces by using the **parallelogram of vectors**.

parallelogram of vectors Method of finding the single **resultant** of two **vectors** by drawing them to scale, separated by the correct angle, and completing the parallelogram of which they are two sides. The diagonal of the parallelogram from that angle represents the **magnitude** and direction of the resultant vector.

parameter Variable in terms of which other interrelated variables can be expressed. It is constant for a given situation. *See* **equations of motion**.

parametric equation Equation where the coordinates are individually written as functions or parameters. *Example*: $x = a \cos \theta$ and $y = a \sin \theta$ are equations for a circle with radius a with their centre at 0. θ is the **parameter**, i.e. the angle the radius makes with the x-axis.

partial derivative Derivative of a **function** with respect to one of its **variables**, all other variables in the function being taken as constant. *Example*: let $a = f(x, y, z)$. If a is a differentiable function of x the partial derivative is written as $\delta a/\delta x$. So, if $a = 3x^2y^3$:

$$\frac{\delta a}{\delta x} = 6xy^3$$

and

$$\frac{\delta a}{\delta y} = 9x^2y^2$$

partial fraction One of the component fractions into which another fraction can be separated, so that the sum of the partial fractions equals the original fraction. *Example*:

$$\frac{2}{x+4} \quad \text{and} \quad \frac{3}{2x-3}$$

are added

$$\frac{2}{x+4} + \frac{3}{2x-3} = \frac{2(2x-3)+3(x+4)}{(x+4)(2x-3)}$$

$$= \frac{7x+6}{(x+4)(2x-3)}$$

Pascal's triangle Triangular pattern of numbers in which each number is the sum of the two above it. It begins:

```
            1
          1   1
        1   2   1
      1   3   3   1
    1   4   6   4   1
  1   5  10  10   5   1
1   6  15  20  15   6   1
```

Each row of numbers gives the **coefficients** of the expansion of the **binomial** $(1+x)^n$. The triangle can therefore be used to calculate **combinations** in **probability**. *See*: **binomial expansion**; Appendix III.

PC Abbreviation for personal computer, now generally used to describe a desktop computer (i.e. one which will fit on the top of an average-sized desk).

pentagon Five-sided plane figure or **polygon**. Each of the interior angles of a regular pentagon equals 108°. *See* **regular polygon**.

per annum (p.a.) Latin phrase meaning 'annually'.

■ **percentage** Fraction expressed in hundredths (with the denominator omitted). [2/4/m] *Example*: one quarter $=1/4=25/100=25$ per cent (often written 25%).

■ **percentile** Hundredth part of a range of statistics (data) of equal frequency. [5/7/c] *Example*: the 80th percentile is the value below which 80 per cent of all the values fall. The 50th percentile is the **median**. *See also* **quartile**.

perfect number Integer that is equal to the sum of its factors, excluding itself but including 1. If n is an integer, then $2^{n-1}(2^n-1)$ is a perfect number. *Example*: if $n=3$, then

$$2^{3-1}(2^3-1)=2^2\times 7=4\times 7=28$$

perfect square Product of two identical numbers. *Example*: $100=10\times 10$; $81=9\times 9$.

perigee Position on the orbit of a moon or artificial satellite at which it is closest to the planet it is orbiting.

perihelion Position on the Earth's orbit at which the planet is closest to the Sun. *Compare* **aphelion**.

■ **perimeter** Distance round the boundary of a **plane** figure, in a **polygon** equal to the lengths of all the sides added together. *Example*: the perimeter of a **rectangle** 3 m by 2 m is $3+3+2+2=10$ m. The perimeter of a circle is called its

circumference, and it is equal to $2\pi r$, where r is the radius of the circle. [4/4/f]

periodic function Function that returns to the same value at regular intervals.

peripheral unit Equipment that can be linked to a **computer**, including input, output and storage devices.

permutation Number of ways a set of numbers can be arranged and ordered. For n numbers, there are $n!$ ways of arranging them n at a time, and ${}^nP_r$, or

$$\frac{n!}{(n-r)!}$$

ways of arranging them r at a time ($n!$ stands for **factorial** n). *Example*: there are $3! = 6$ ways of arranging the three numbers 678 three at a time: 876, 768, 678, 867, 786 and 687. Most **scientific calculators** have a ${}^nP_r$ key which will calculate the permutations of arranging n numbers r at a time (*see* Appendix II).

perpendicular Describing a line or plane that is at right angles to another line or plane. *See also* **vertical**; **normal**.

personal computer (PC) *See* **microcomputer**.

perturbation theory Method of obtaining approximate solutions to equations representing the behaviour of a system. It is used in quantum mechanics.

peta- (P) Prefix in the **metric system** signifying $\times 10^{15}$.

■**pi** (π) Name given in mathematics to: *1.* the **ratio** between the **circumference** of a **circle** and its **diameter**. It is an **irrational number** with the value 3·1415926536 . . . 'Pi' is the name of the Greek letter for 'p', and the symbol π is the Greek letter pi. [2/9/a; 4/5/h] *2.* In **radian** measure, π is used to signify 180°.

pico- (p) Prefix in the **metric system** signifying $\times 10^{-12}$.

■ **pictogram** Method of graphically representing statistical data that uses symbols, each representing a given number (of items of information). [5/3/c] *Example*: if a barrel is used to symbolize one million gallons of oil, a row of 7 barrels represents 7 million gallons. It is thus a type of pictorial **bar chart**.

■ **pie chart** Method of graphically representing statistical data that consists of a circular diagram divided into sectors whose size (determined by the angle they subtend at the centre) represents a number (of items of information), calculated as a percentage of the whole. [5/5/d] *Example*: an item allocated a quadrant (quarter of the pie chart) represents 25 per cent of the whole.

■ **pint** (pt) Unit measure of liquid capacity equal to one-eighth of a gallon = 0·57 **litres**. [2/5/h]

place-value notation System of arranging numbers so that their position in relation to each other gives each number a particular value. *Example*: in **denary**, numbers are horizontally arranged from right to left according to the values of units, tens, hundreds, thousands, and so on. Therefore, when we see the denary expression 45567, we read it as $4 \times 10^4 + 5 \times 10^3 + 5 \times 10^2 + 6 \times 10^1 + 7$ units. *See also* **binary notation**.

plan *1.* Projection of an object on a horizontal plane, usually drawn to scale, and showing a relative position of objects. *2.* Intended way of carrying out a mathematical process.

planar Describing something that is flat or occupies a plane.

plane Flat surface, imaginary or real. In **coordinate geometry**, a plane has the general equation:

$$ax + by + cz + d = 0$$

■**plane of symmetry** Plane that divides a solid figure into two equal parts. In a **sphere** this would form a plane through the centre of the sphere. [4/3/b]

point *1.* Position at which two lines meet, also termed a single position of intersection. *2.* Dot used to distinguish the integral part of a **decimal** number from its fractional part. Alternative term: **decimal point**.

■**point symmetry** Point equidistant from opposite sides of a figure, e.g. P in Figure P1. [4/4/e]

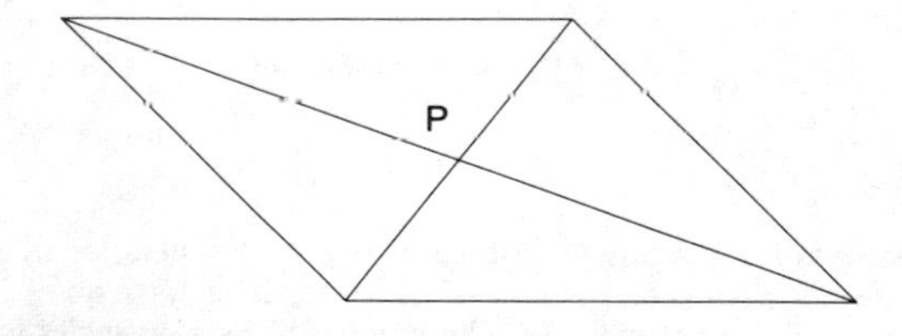

Fig. P1 *point symmetry*

polar coordinates System of fixing the coordinates of a point with reference to a fixed point and the angle this line makes with an initial fixed line, as shown in Figure P2, where the point P is described by the length of r and the size of θ, and $x=r \cos \theta$, $y=r \sin \theta$. It is one of the systems used to describe the location of a point in **coordinate geometry**. Most **scientific calculators** will perform conversions from polar coordinates to rectangular coordinates (i.e. **Cartesian coordinates**), and vice-versa. *See* **bearing**.

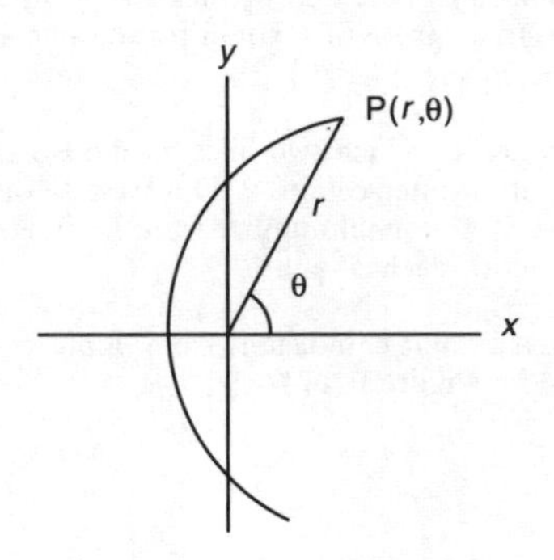

Fig. P2 *polar coordinates*

polygon Plane figure with three or more sides, none of which intersect each other, but all of which together form an enclosed geometric figure. The number of **interior angles** in any polygon is equal to the number of sides as in, for example, a **hexagon** or **octagon**. The sum of the interior angles of any polygon is equal to $(n-2)\times 180°$, where n is the number of sides. The sum of the exterior angles of any polygon is 360°, and a polygon has $\frac{1}{2}n(n-3)$ diagonals. *Example*: the sum of the interior angles of a **heptagon** equals $(7-2)\times 180° = 900°$; a heptagon has $3{\cdot}5(4) = 14$ diagonals. *See also* **regular polygon**.

polyhedra Plural of **polyhedron**.

polyhedron (plural polyhedra) **Solid** figure with many **faces**. There are five types of **regular polyhedron**, as follows:

1. tetrahedron (4 triangular faces);
2. cube (6 square faces);
3. octahedron (8 triangular faces);
4. dodecahedron (12 pentagonal faces);
5. icosahedron (20 triangular faces).

polynomial Algebraic expression with only one **variable**. *Example*:

$$a_0x^n + a_1x^{n-1} + a_2x^{n-2} \ldots a_{n-1}x + a_n$$

population Term applied in statistical analysis to the set of data which all contain similar properties.

position circle Circle with its centre at an observed point and its radius such that the circumference passes through the place of observation.

position line Line of position at which the observer is situated at a given time.

positive Describing a number or quantity greater than zero, denoted by the sign +. A number can be taken as positive unless it is preceded by the − sign.

pound (lb) Unit of weight on the **f.p.s. system**, equal to 16 **ounces** = 0·454 kg.

■ **power** *1.* Rate of doing work in mechanics. Measured in watts. *2.* Number used to denote how many times a quantity has been multiplied by itself. *Example*: $2 \times 2 \times 2 \times 2$ could be written as 2 to the power of 4 (2^4) = 16. [2/5/k] [2/8/a] [3/7/c]

pr Abbreviation of **probability, mathematical**.

prime Alternative term for **prime number**.

■ **prime number** Integer (whole number) divisible only by itself and 1, e.g. 2, 3, 7, 11, 13, 17, 19 . . . ; the largest prime number so far discovered (calculated on a **super computer**) is $2^{216091} - 1$. All prime numbers (or primes) except 2 are **odd numbers**. Alternative term: prime. [3/5/c] *See also* **sieve**.

prime symbol Symbol ′, used to distinguish pairs of terms where one is a development of the other, for example in a **centre of enlargement** where one geometric figure is enlarged to create a second with the same named vertices, the first set of vertices is denoted as prime. The symbol is also used to denote **foot, minute** and **recurring decimal**.

principal (P) Sum invested on which interest is payable. *See* **annual percentage rate**; **compound interest**; **simple interest**.

printer Computer **output device** that produces hard copy (i.e. output on paper) as a printout. There are various kinds, including (in order of speed) daisy-wheel, dot-matrix, line, barrel and laser.

printout Output (**hard copy**) from a computer **printer**.

■**prism Polyhedron** of two congruent **polygon** bases with sides parallel and **vertices** joined by parallel lines. A **cube** is a **rectangular** prism. [4/2/a]

■**probability, mathematical** (pr) Mathematical expression of the extent to which an event is likely to occur, given a value between 0 (an impossibility) and 1 (a certainty). [5/4/g,h,i] *Example*: the probability of getting a 6 on one roll of a dice is 1/6 or 0·16666′.

produce In geometry, to extend a given line in the same plane as the original line.

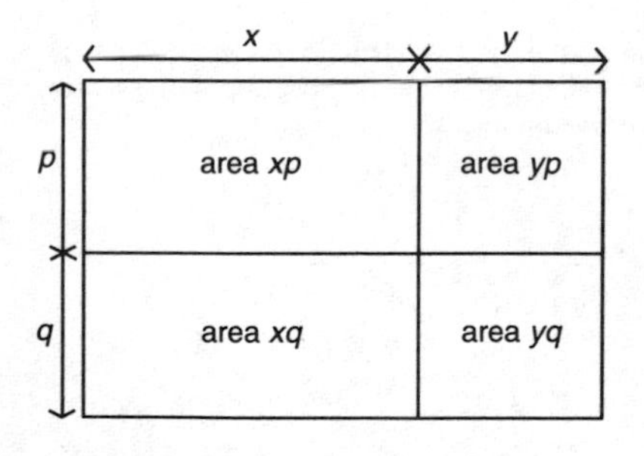

Fig. P3 *product (1)*

■**product** In mathematics, the result of multiplying numbers together. [2/7/c] *Example*: product of 2 binomial expressions $x + y$ and $p + q$. If we want to multiply these together, Figure P3 could be considered.

$$\text{The area of the whole} = (x+y)(p+q)$$
$$= \text{sum of individual areas}$$
$$= xp + xq + yp + yq$$

If both sets of terms are looked at as in Figure P4, a pattern emerges:

$$= xp + xq + yp + yq$$

The terms on the right-hand side can be obtained by multiplying the terms of the left-hand side, as shown in Figure P5

$$= 2x^2 + 8x + 3x + 12$$
$$= 2x^2 + 11x + 12$$

program Sequence of instructions for a **computer**, or for a **scientific calculator** which is programmable. Alternative name: programme.

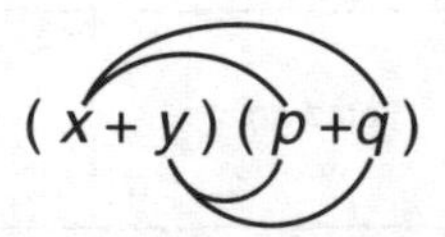

Fig. P4 *product* (2)

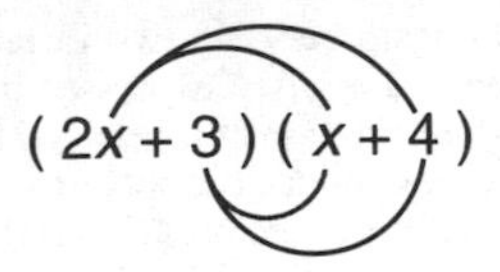

Fig. P5 *product (3)*

progression Mathematical **series** of terms each of which is linked by a common operation. *Example*: 12, 16, 20, 24, 28, 32, 36 is an arithmetic progression in which 4 is added to the each number to give the next. In a progression such as 1, 4, 16, 64, 256 each number is multiplied by 4 to give the next. This is a **geometric progression**.

projectile Missile that is capable of being projected by means of a force.

projection Transformation in which a shape or set of points is mapped onto a plane. It may be a one-to-one transformation where the original shape is maintained, or a distortion may take placc.

proof *1.* Logical statement using evidence that helps to establish a fact or hypothesis. *2.* Argument put forward, sometimes using experimental or empirical observations, to establish the validity of certain relationships.

proper fraction Fraction whose **numerator** is less than its **denominator**, e.g. $\frac{5}{6}$, $\frac{11}{16}$, $\frac{2}{3}$. *See also* **improper fraction**; **vulgar fraction**.

proper subset Any **set** between two extremes of the **null set** and the original set.

proportion Relationship between two **variables**, which may involve a **function** and is often expressed in the form of an equality. A constant relationship between the two variables (e.g. $x \propto y$) is termed direct proportion. Where the relationship of one of the reciprocal variables to the other variable is constant, e.g. $x \propto 1/y$, the term inverse proportion is used. *See also* **golden section**; **ratio**.

■**protractor** Instrument used for measuring or constructing angles. It is usually semicircular in shape and divided into 180°. [4/4/c]

pt Abbreviation of **pint**.

punched card Computer input or output medium consisting of cards punched with coded holes, now very much outdated. The actual input device used was a punched card reader; the output device was a card punch.

punched tape Computer input or output medium consisting of paper tape punched with coded holes, now outdated. The input device was a punched tape reader; the output device was a tape punch.

pyramid Solid with a **polygon** as a base and three or more triangular sides meeting at an apex. Its volume equals the product of $\frac{1}{3}$ of the area of the base and the perpendicular height. *Example*: if a pyramid is formed with a square with sides 3 m in length, and a height of 8 m, then its volume equals $\frac{9}{3} \times 8 = 24$ cubic metres (m^3).

■**Pythagoras, theorem of** Theorem which states that in a right-angled triangle, the square of the length of the hypotenuse (longest side) is equal to the sum of the squares of the lengths of the other two sides. Alternative name: Pythagorean theorem. It was named after the Greek philosopher and mathematician Pythagoras (*fl. c.* 530 BC). [4/7/c] *See* Appendix III. The theorem can easily be proved. *Example*: construct a right-angled triangle with sides (for example) approx 50 mm, 50 mm and 70 mm. Using each side of the triangle as a base, construct squares on all three sides, as in Figure P6. Calculate the area of each of the three squares: $50 \times 50 = 2500$ mm; $50 \times 50 = 2500$; $70 \times 70 = 4900$. Allowing room for error, the area of the square on the longest side is equal to the sum of the areas of the squares on the other two sides.

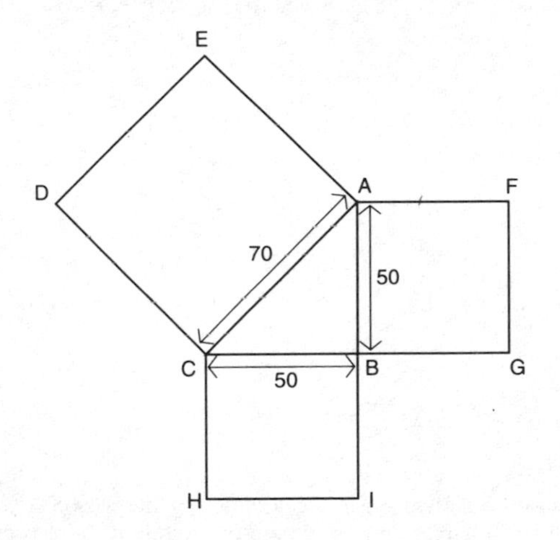

Fig. P6 *Pythagoras' theorem*

Q

QED Abbreviation for ***quod erat demonstrandum***.

QEF Abbreviation for ***quod erat faciendum***.

qt Abbreviation for **quart**.

■**quadrant** Quarter of a circle, as shown in Figure Q1, where the shaded area of the circle is the quadrant. [4/4/d]

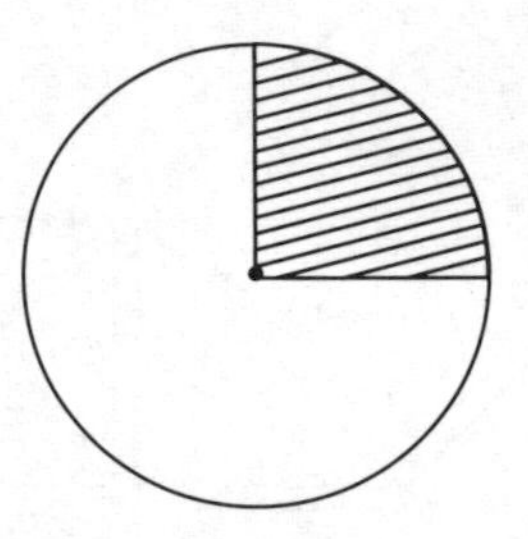

Fig. Q1 *quadrant*

■**quadratic equation** Algebraic equation of the second order (degree) or square power. Quadratic equations have two possible solutions; these are termed the **roots** of the unknown. The roots may be real and different, real and the same (coincident) or **imaginary**. [3/10/b] The general quadratic equation (**1**) is

$$ax^2+bx+c=0$$

and the solutions are given by (**2**)

$$x=\frac{-b\pm(\sqrt{b^2-4ac})}{2a}$$

Example: solve $x^2-6x+4=0$. If this equation is compared with the general quadratic equation (**1**) above, then it can be seen that $a=1$, $b=-6$, and $c=4$. If these values are substituted in the solution equation (**2**), you have

$$x=\frac{-(-6)\pm\sqrt{(-6)^2-4\,.\,1\,.\,4}}{2\,.\,1}$$

$$=\frac{6\pm\sqrt{36-16}}{2}$$

$$=\frac{6\pm\sqrt{20}}{2}$$

$$=\frac{6\pm4{\cdot}47}{2}$$

$$=\frac{6+4{\cdot}47}{2} \quad \text{or} \quad \frac{6-4{\cdot}47}{2}$$

$$=\frac{10{\cdot}47}{2} \quad \text{or} \quad \frac{1{\cdot}53}{2}$$

$$=5{\cdot}24 \quad \text{or} \quad 0{\cdot}77$$

See also **completing the square**.

■ **quadrilateral** Four-sided plane figure. A **square**, **rectangle**, parallelogram, kite, **rhombus** and trapezium are all examples of a quadrilateral. [4/6/d]

quart Unit of fluid measure. One quart = 2 **pints** = approx 1·14 **litres**; 4 quarts = 1 **gallon**.

quartic Having the property of 4. A quartic equation is an equation of the fourth order, for example

$$x^4+3x^2+6x-3=0$$

A quartic expression is one in the fourth degree.

■ **quartile** In statistics, the value below which one-quarter of a set of data lies, and the value above which three quarters of the set of data lies; the 25th and 75th **percentiles**, and the first and third quartiles, respectively. The second quartile equals the **median** of the set of data. [5/8/c] *See also* **interquartile range**.

quindecagon **Polygon** which has 15 sides.

quintic Having the property of 5. A quintic equation is an equation of the fifth order. A quintic expression is one in the fifth degree.

quinquagenarian Person reaching the ages between 50 and 60.

quod erat demonstrandum (QED) Latin phrase meaning 'that which is to be proved', most often used in the abbreviated form 'QED'. The term is used when the proof of a mathematical statement has been fully demonstrated. *Example*: a person asked to prove that a given **polygon** is a **pentagon** could, after showing that the sum of the interior angles equalled 540°, write QED at the end of the working out, because the sum of the interior angles of a polygon equalling 540° is a **necessary condition** of its being a pentagon. *Compare* ***quod erat faciendum***.

quod erat faciendum (QEF) Latin phrase meaning 'that which was to be done', most often used in the abbreviated form 'QEF'. The term is used when a given task has been completed. *Example*: a person asked to construct a **rhombus** of a given area could, having done so, write QEF below the completed figure. If asked to prove the figure is a rhombus then, having written the proof, the person can write **QED**.

quotient Result of **division**. For example, in $35 \div 7 = 5$, the number 5 is the quotient; in $23 \div 7 = 3$, remainder 2, the number 3 is the quotient.

qwerty keyboard Description of the standard **alphanumeric** keyboard used on typewriters, typesetting machines, word processors and computers (named after the first six letters on the top rank of letter keys).

R

R Abbreviation for **rate** (sense *2.*).

r Abbreviation for **radius**.

rad Abbreviation for **radian**.

■**radial symmetry** Symmetry about any one of several lines or planes through the centre of an object or organism. [4/4/e] *See also* **bilateral symmetry**; **rotational symmetry**.

radial velocity Alternative name for **line-of-sight velocity**.

radian (rad) **SI unit** of plane angle; the angle at the centre of a **circle** subtended by an **arc** whose length is equal to the **radius** of the circle. Since the circumference of a circle $= 360° = \pi \times \text{diameter} = \pi 2r$, the circumference of a semicircle $= 180° = \pi r$. Therefore to express the radius of the circle as a value in degrees, 1 radian $= 180° \div \pi = 57{\cdot}2957°$ or, to put it another way, $180° = \pi \times 1$ radian. It follows from the last statement that there are $2\pi \times$ radians in every circle (if $180° = \pi$ radians, then $360° = 2\pi$ radians). To convert degrees to radians, use the formula

$$\text{rad } \theta = \theta \times \frac{\pi}{180}$$

Example: $\text{rad } 65° = 65 \times \dfrac{\pi}{180}$

$$= 65 \times 0{\cdot}01745$$

$$= 1{\cdot}13446$$

To convert radians to degrees, multiply the radian value by $\frac{180}{\pi}$. *Example*: convert 0·45378; 0·73303 and 1·65806 to degrees

$$0{\cdot}45378 \times \frac{180}{\pi} = 26°$$

$$0{\cdot}73303 \times \frac{180}{\pi} = 42°$$

$$1{\cdot}65806 \times \frac{180}{\pi} = 95°$$

To save the inconvenience of performing the calculations on every occasion, tables can be consulted giving the radian value of degrees from 0° to 360° and giving the degree values of radians from 0·1 to 10. Some calculators also have keys which can convert from radians to degrees and vice-versa. An alternative way to express radian values can be derived from the formula 180° = π radians. This is often expressed in the abbreviated form 180° = π (the term 'radian' is understood as being present). This allows the use of expressions such as $\frac{\pi}{2}$; $\frac{\pi}{6}$ and $\frac{\pi}{4}$ which equal 1·57079, 0·52359 and 0·78539 in radians respectively. Thus when an expression such as

$$\cos\frac{\pi}{7} = 0{\cdot}14285$$

is used, it should be understood that the cosine of a radian value is being found. Use of π in this context should not be confused with its other application, i.e. as the ratio of the circumference to the diameter of a circle = 3·14159. Because radian measure is so closely linked to the relationship between the circumference and radius of a circle, a value supplied in radian measure makes it possible to calculate the length of the arc of any circle subtended by a given radian using the formula

$$l = \theta r$$

where l = length of the arc and r = the radius of the circle. *Example*: calculate the length of arc subtended by an angle of 73° in a circle of radius 45 cm. First find rad 73°

$$\text{rad } 73° = 73 \times 0{\cdot}017453 = 1{\cdot}27409$$

Now the rad is known, the formula can be followed, giving

$$\begin{aligned} l &= 45 \times 1{\cdot}27409 \\ &= 57{\cdot}33406 \text{ cm} \end{aligned}$$

From this it is evident that radians are, in effect, the **ratio** in any circle of the length of arc to the length of the radius: radian 1 (57·29577°) is the 1 to 1 ratio of length of arc to radius. Since the circumference of a circle = $\pi 2r$, it follows, therefore, that radian 2 will subtend a length of arc = $2r$ and that radian 6 will subtend an arc almost equal to the complete circumference of any circle. Since the total area of a circle can be represented by the formula πr^2 and by the term 2π, and the ratio of any angle to a complete circle = $\frac{\theta}{2\pi}$, then the area of any **sector** of a circle subtended by an angle θ can be expressed as follows

$$\frac{\text{area of sector}}{\pi r^2} = \frac{\theta}{2\pi}$$

Therefore

$$\text{area of sector} = \tfrac{1}{2} r^2 \theta$$

Example: in a circle with radius 6 m, what is the area of a sector subtended by an angle of $\frac{\pi}{4}$?

$$\begin{aligned} A &= \tfrac{1}{2}r^2\theta \\ &= \tfrac{1}{2}6^2 \times \frac{\pi}{4} \\ &= \tfrac{1}{2}36 \times 0{\cdot}78539 \\ &= 18 \times 0{\cdot}78539 \\ &= 14{\cdot}137\ \text{m}^2 \end{aligned}$$

Use of radian measure considerably simplifies calculations using circular angles, and has important applications in **calculus**.

radical In mathematics, relating to the **root** of a number or quantity. The symbol $\surd$ is known as the radical sign.

radii Plural of **radius**.

radius (r) Distance from the centre of a **circle** to the **circumference**, equal to half the **diameter** or to the circumference divided by 2π. Plural: radii.

radius of curvature Of a point on a curve, the **radius** of a **circle** that touches the inside of the curve at that point.

radius of gyration Of a rotating object, the distance from the **axis** of rotation at which the total mass of the object might be concentrated without changing its **moment of inertia**.

radius vector Distance from the **origin** to the point in a plane with the **polar coordinates** (r and θ).

radix *1.* **Base** number of a counting system; e.g. 10 is the radix of everyday numbers, whereas 2 is the radix of **binary notation**. *2.* A **root** of a number.

RAM Abbreviation of **random access memory** of a **computer**.

random Describing a sample of **raw data**, i.e. a collection of unordered **variables**.

random access memory (RAM) Part of a computer's **memory** that can be written to and read from. *See also* **ROM**.

range Interval between the smallest and largest value in a set of data. The range of a **variable** could be all values of this in a given interval.

rank Quality of a **determinant** in a matrix, and thus a quality of a matrix itself. If the matrix has a rank r then it contains at least one determinant of order r and all higher order determinants are zero.

■**rate** *1.* Relation between two **variables** in which one of the variables is time, e.g. speed is the ratio of distance divided by time. *2.* (R) Percentage **interest** charge made on sums of money borrowed, or paid on sums of money invested, e.g. an interest rate of 8%. [3/9/e]

■**ratio** Pair of numbers or terms that represents a relationship between the two numbers or terms, often expressed as a **proportion** or **fraction**. [2/5/d] *Example*: if there are 4 black objects and 6 white objects, the ratio of black objects to white ones is 4 to 6 (written 4:6) which, like the fraction $\frac{4}{6}$, can be reduced to 2:3. Note, however, that not all fractions are ratios, or vice versa. Ratios are normally signified by the : symbol placed between the two numbers which form the ratio. Ratios between the sides of geometric figures are usually represented as, for example $\frac{AB}{BC}$ *See also* **cosine**; **golden section**; **sine**; **tangent**; **trigonometrical ratio**.

rationalize Transformation of equations which contain irrational numbers in order to simplify them. Fractions can also be simplified using this method. *Example*:

$$\sqrt{6x-3}-x=1$$
$$\sqrt{6x-3}=1+x$$
$$6x-3=(1+x)^2$$
$$6x-3=1+2x+x^2$$
$$x^2-4x-4=0$$

■**rational number** Number that may be expressed as a ratio of two integers, i.e. in the form of a **fraction**. [2/9/a] *See also* **irrational number**.

raw data Set of data that has not been ordered or arranged in any way. It is exactly as it has been collected.

ray *1.* In **projections**, a straight line originating from a point. *2.* In **vector** notation, a line which has direction. *Example*: a line representing the direction from X to Y would be written as XY.

read head Electromagnetic device on a tape recorder, video recorder or computer that 'reads' signals stored on a magnetic tape, disk or drum.

read-only memory (ROM) Part of a computer's **memory** that can only be read (and not written to). *See also* **random access memory**.

read-write head Electromagnetic device that functions as both a **read head** and a **write head**.

real number Any number, positive or negative, from among all **rational numbers** and **irrational numbers** (as opposed to an **imaginary number**). It could be the real part of a complex number, e.g. $a+ib$ where a is the real part and i the imaginary part.

■**reciprocal** Quantity obtained by dividing a number into 1; i.e. the reciprocal of x is the number $1/x$. Zero has no reciprocal. Most **scientific calculators** have a reciprocal key, marked $\frac{1}{x}$ or x^{-1}. *Example*: the reciprocal of 26 is $1 \div 26 = 0{\cdot}038$ (on a calculator, key 26, then reciprocal, then =); the reciprocal of $0{\cdot}50$ is $1 \div 0{\cdot}50 = 2$; the reciprocal of $\frac{6}{5} = \frac{1}{6/5} = \frac{5}{6}$. [3/7/d]

■**rectangle** Plane figure bounded by four straight sides, with opposite sides equal and all of the sides at right angles to each other. Its area is the product of the lengths of two adjacent sides. [4/2/a]

rectangle number Number that can be represented by a rectangular pattern of dots, e.g. 18 can be represented by:

. . . . . .

. . . . . .

. . . . . .

rectangular Right-angled: the lines which form a **square** are rectangular.

rectangular hyperbola Hyperbola that has the x-axis and y-axis as its **asymptotes** (general equation $xy=c^2$).

rectilinear In, or forming, a straight line.

recurring decimal Decimal that contains a number or block of numbers that repeat to infinity. *Example*: the decimal for $\frac{2}{3}=0{\cdot}666666\ldots$; for $\frac{2}{11}=0{\cdot}181818\ldots$ Recurring decimals are often signified by the the use of **prime symbol** ′, as in 0·666′ as well as by the use of repeated full stops (. . .). *See also* **significant figure**; **terminal decimal**.

reduction Operation of reducing a **fraction** to its simplest form, usually by finding the **lowest common denominator**. *Example*: $\frac{6}{12}=\frac{1}{2}$, because the LCD is 6; $\frac{23}{3}=7\frac{2}{3}$. Some fractions can only be reduced to a limited extent: $\frac{85}{23}=3\frac{16}{23}$; in cases like this, it can be easier to express the remainder as a decimal, i.e. 3·69565.

re-entrant polygon Many-sided plane figure with an internal angle larger than 180°.

■ **reflection Transformation** in geometry when a point (P) has an image (P′) the same distance behind a plane as the point is in front, as shown in Figure R1, where the line of reflection is the perpendicular bisector of PP′. [4/3/b]

reflex angle In geometry, an angle that exceeds 180°, but is less than 360°.

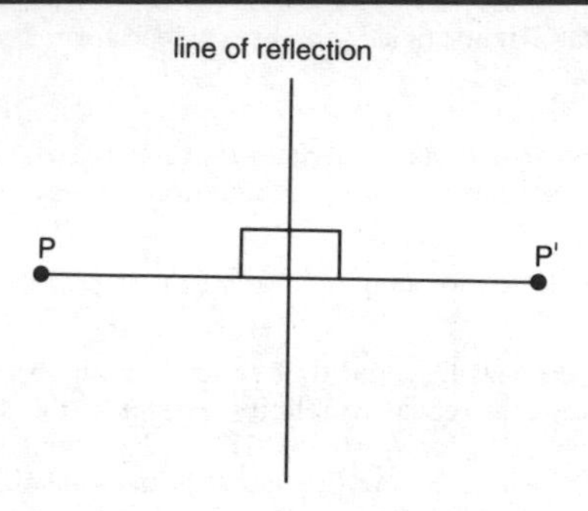

Fig. R1 *reflection*

reflexive Quality of a mathematical relationship which represents equality, e.g. $y = y$ for all values of y.

region *1.* Space limited by precise boundaries. *2.* Set of points, alternative term **domain**. *3.* Set of values. *4.* Independent **variable** for which a certain **function** is defined.

regular polygon Many-sided **plane** figure with all its sides and internal angles equal, for example a **pentagon**, **hexagon** or **square**. Each of the **interior angles** in a polygon is equal to

$$\frac{(n-2) \times 180}{n}$$

where n equals the number of sides in the polygon. The exterior angles of a regular polygon are equal to $\frac{360°}{n}$; the sum of the exterior angles is, therefore, equal to 360°. *See* **polygon**.

regular polyhedron Solid figure with all its faces the same regular shape (all regular **polygons**). There are five regular polyhedra. *See* **polyhedron**.

relation Connection between two terms or values. Quantities are related often by algebraic expressions.

relativity Einstein's theory that time passes relatively according to the speed at which two different observers are moving, and that the mass of a body varies according to its motion. These differences only become apparent at very high speeds, i.e. those approaching the speed of light ($2{\cdot}997 \times 10^8$ m s^{-1}). Einstein developed two theories of relativity: the General Theory, and the Special Theory. *See* Appendix II.

remainder Number left over during division if the **quotient** (or result) is not an integral value, e.g. if 20 is divided by 3 the answer is 6, and the remainder is 2.

remainder theorem Theorem enabling the remainder to be found when a **polynomial** expression is divided. When a polynomial is divided by $(x-a)$ the remainder is found by writing a for x in the original expression. *Example*: find the remainder when

$$4x^3+7x^2+6x-2$$

is divided by $(x-3)$. Substitute $x=3$, which gives

$$\begin{aligned}4\times3^3+7\times3^2+6\times3-2&=4\times27+7\times9+18-2\\&=108+63+18-2\\&=187\end{aligned}$$

repeated root If a function $f(x)$ can be divided by $(x-a)^n$, then if the function $f(x)$ is equated to zero, a will be the root n times. *Example*:

$$x^3+4x^2-3x-18=0$$

factorizes to $(x-2)(x+3)^2$. Therefore 2 is a normal root and -3 is a double root. If this was plotted, the curve would pass once through $x=2$ and twice through $x=-3$.

residue In **modulo arithmetic**, the equivalence classes for an integer n of integers partitioned by their remainders when they are divided by n (a given positive integer). *Example*: 16 is in residue class 2 when dividing by 7.

■**resolution** Obtaining two **vectors** which are equivalent to an original vector. The single vector is resolved into two **components** by constructing a **parallelogram** where the original vector is represented by the diagonal and, due to the **parallelogram of vectors** rule, the two component vectors are represented in magnitude and direction by the two adjacent sides of the parallelogram. *Example*: in Figure R2, the two component vectors are R sin θ and R cos θ. [4/9/e]

■**resultant** Of two or more **vectors**, the single vector that has the same effect. It can be found by constructing the **parallelogram of vectors**. [4/9/e]

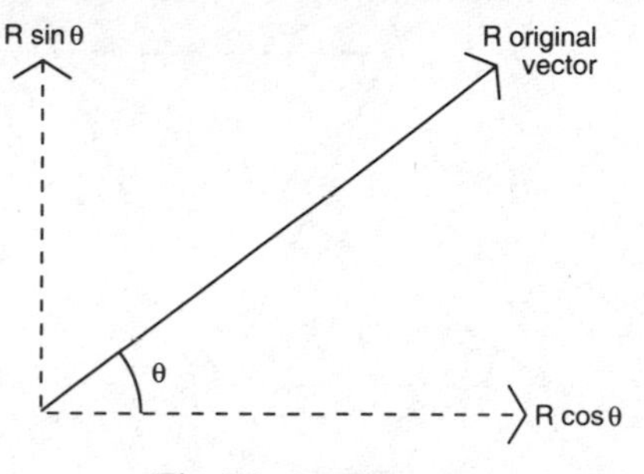

Fig. R2 *resolution*

revolve To rotate about an **axis** or point. One complete revolution would be 360° or 2π radians.

■**rhombus Parallelogram** with all of its sides equal. Its diagonals bisect each other at right angles, as shown in Figure R3. [4/2/a]

■**right angle** Angle equal to one quarter of revolution, or 90°. [4/4/c]

■**right-angled triangle** Triangle that includes as one of its three angles a right angle (angle equal to 90°). It therefore follows that the other two angles must be **complementary** (add up to 90°). The side of the triangle opposite the right angle is

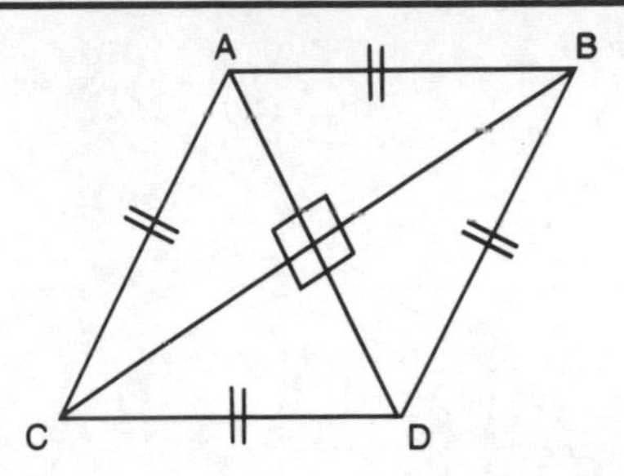

Fig. R3 *rhombus*

termed the **hypotenuse**, i.e. the square of the length of this side is equal to the sum of the squares of the lengths of the other two sides, as stated in the theorem of **Pythagoras**. In any right-angled triangle, the **altitude**, i.e. the line dropped from the vertex containing the right angle to meet the hypotenuse, is a proportional **mean**. This is shown in Figure R4, where

$$\frac{BD}{AD} \text{ as } \frac{AD}{DC}$$

In any right-angled triangle, the three sides are in a constant proportion to one another, and these proportions are expressed in the **trigonometric ratios**. In any right-angled triangle, the hypotenuse is the **diameter** of a **circumscribed circle** around the triangle (as also shown in Figure R4). [4/2/a]

rms Abbreviation for **root mean square**.

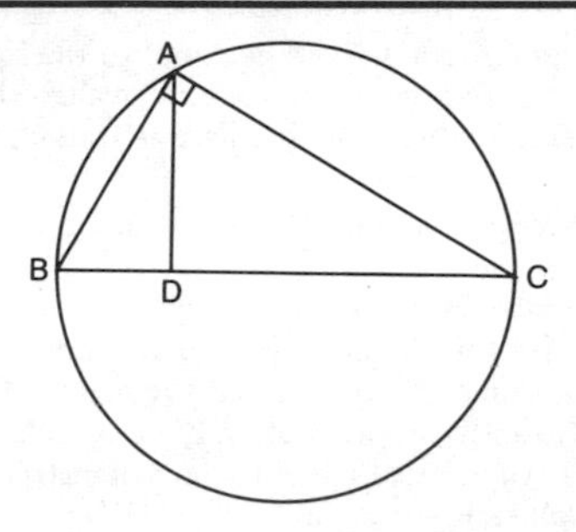

Fig. R4 *right-angled triangle*

ROM Abbreviation of **read-only memory** of a **computer**.

Roman numerals Number system, originally used by the Romans, based on letters: I = 1, V = 5, X = 10, L = 50, C = 100, D = 500, M = 1,000. Other numbers are written using combinations of these: IX = 9, XL = 40, XC = 90, CD = 400, CM = 900. There is no zero.

root *1.* Number or quantity that, when multiplied by itself a number of specified times, gives another number; e.g. the **square root** of 9 is 3; the cube root of 27 is 3. Most **scientific calculators** have both a square root key (marked $\sqrt{}$) and a general root key (marked $\sqrt[x]{}$) which gives the *x*th root of a number; some have a cube root key in addition to these. *2.* Solution of an algebraic **equation**.

root mean square (rms) Average equal to the **square root** of the sum of the squares of a number of values divided by the total number of values. *See* **standard deviation**; **variance**.

rotation Turning all points about a plane, point or line.

■**rotational symmetry** In geometry, symmetry of a figure when it is rotated round a point. Rotational symmetry is given an order, e.g. a square has rotational symmetry of order 4 since its shape is exactly repeated after 1/4 of a rotation: $360°/4 = 90°$. [4/4/e] Alternative term: **radial symmetry**. *See also* **bilateral symmetry**.

rounding error Reduction or truncation of numbers, particularly decimal numbers. This obviously introduces an error, e.g. 3·1415 could be rounded to 3·141 or 3·142. The number is rounded to the nearest whole number in order to minimise the possible error. In this instance, 3·142 is the more accurate rounding because 0·0005 follows ·3141; where a decimal number is followed by a 5 or greater number, round up the preceding number to the next greatest number; where it is 4 or less, leave the preceding number as it is. *See also* **significant figure**.

rounding off Approximation of numbers to allow a rough estimate for a mathematical process. It is one method of getting a quick result from a calculation, e.g. $2{\cdot}875 \times 1{\cdot}135$ could be rounded off to $3 \times 1 = 3$ (the actual number is 3·263125). *See also* **accuracy**; **estimate**; **significant figure**.

row Number of mathematical quantities arranged horizontally in a straight line.

■**row matrix** Matrix which only contains elements arranged in a row, e.g. (1 2). [4/10/d]

S

S *1.* Abbreviation of **set**. *2.* Abbreviation of **sum**. *3.* Abbreviation of **compass** direction South.

s Symbol for distance. *See* **equations of motion**.

s Abbreviation of **second**.

salary Money paid to an employee, usually expressed as an annual figure, and paid directly into the employee's bank account. *See* **gross**; **net**. *Compare* **wage**.

sample In statistics, data chosen to represent a **population** chosen for study. Often the data is recorded as it occurs, hence producing a **random** sample.

satisfy *1.* To fulfil the conditions of a mathematical operation by producing a result which occurs as part of the operation. *2.* To meet the **proof** of a mathematical statement adequately. *See also* ***quod erat demonstrandum***.

savings account Bank, building society or other account into which the holder puts money on condition (for example) that sums will only be withdrawn after an agreed notice period. In return, the account holder is paid **interest** on the money held in the account. *See also* **deposit account**.

scalar Of quantities (such as time or weight), having **magnitude** but not direction (unlike a **vector**, which has both).

scalar product **Product** of **vector** quantities, equal to the product of the **magnitudes** of the two vectors and the **cosine** of the angle between them. *Example*: if vectors are A and B

$$A \,.\, B = |A| \,.\, |B| \,.\, \cos \theta$$

where θ is the angle between the vectors. Alternative term: dot product.

■ **scale** *1.* Sequence of numbers in order, as in **directed numbers**. *2.* Set of marks at equal distances apart on a line used for measuring, for example on a ruler. *3.* Comparative **scale** using a certain ratio to form the basis of scale drawings. *Example*: if, in a plan of a building, the scale 1 cm:1 m was used, then a shape 4 cm by 5 cm would represent an actual shape of 4 m by 5 m. [2/5/e]

scalene triangle Triangle in which no two angles or sides are equal to one another. *See also* **equilateral triangle**; **isosceles triangle**.

scales Mechanical or electronic device used for measuring units of weight, such as **ounces** or **grams**. Care should be token to note which unit system is used.

scanner Alternative name for **optical character reader**.

■ **scatter diagram** Diagram showing the **frequency distribution** of a pair of **variables**, as in Figure S1. The dispersion of the data can be seen, and a positive trend determined. [5/6/c] Alternative term: scattergram. *See also* **line of best fit**.

scientific calculator Widely-used term describing an electronic **calculator** with inbuilt mathematical functions such as π, **trigonometrical ratios**, **logarithms**, square and cube **root**,

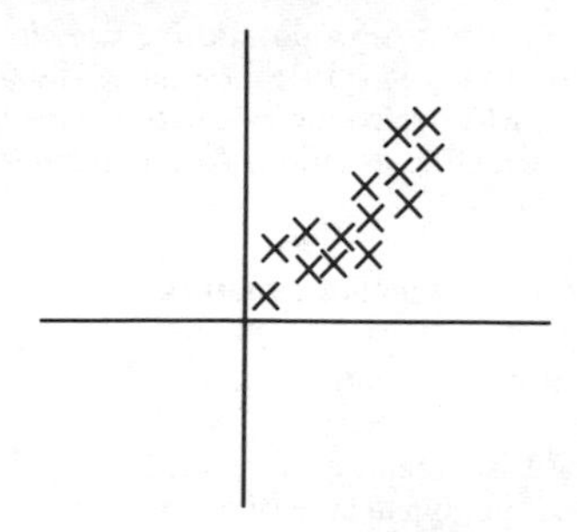

Fig. S1 *scatter diagram*

exponent, **binary**, **hexadecimal** and **octal** bases, and **radian** and **degree** angular measure. Such calculators usually employ a multifunction keyboard controlled by an inverse or 'shift' key determining which of possibly several functions are initiated when pressing a particular key. Depending on the calculator, scientific calculators also have a **program** facility and several levels (and capacities) of memory which will store programs for future use, and a variety of modes (e.g. scientific, engineering, statistical) in which it is possible to use the calculator. *See* Appendix II.

scientific notation System of number representation which expresses numbers as multiples of 10 to a given **power**, particularly useful in dealing with very large and very small numbers. The prefixes **milli-**, **mega-** etc. used in the **metric**

system apply to this type of notation. *Example*: the number 35,600 is written as $3{\cdot}56 \times 10^4$; the number 0·00356 is written as $3{\cdot}56 \times 10^{-3}$. Most **scientific calculators** have a facility for calculating in scientific notation. Alternative term: **standard index form**.

sd Abbreviation of **standard deviation**.

sec Abbreviation of **secant**.

■**secant** *1.* Line that intersects a curve. *2.* (sec) **Trigonometrical ratio** equal to: the **hypotenuse** divided by the **base** of a **right-angled triangle**, i.e. 1/**cosine** θ in a right-angled triangle. Tables exist giving such values; it is also possible to obtain them on a calculator with a 'cos' and reciprocal $\left(\frac{1}{x}\right)$ key. *Example*: if $\theta = 36°$, key 36, followed by 'cos' which will give 0·80901, then key $\frac{1}{x}$, which will give 1·23606. *See also* **cosecant**; **cotangent**. [4/9/a]

sech Abbreviation for the **hyperbolic function** $1/\cosh x$.

■**second** (s) *1.* Unit of angular measurement equal to $\frac{1}{60}$ of a **minute** or $\frac{1}{360}$ of a **degree**, symbol ″. [4/4/c]. *2.* Unit of time on the **f.p.s. system** and on the **metric system**, 60 seconds = 1 minute.

section *1.* Plane that passes through a solid shape, called a plane section. *2.* Drawing to scale of the variations in ground level along a certain line.

■**sector** Part of a **circle** between two **radii**. If the radii are at right angles, the sector is a **quadrant**. [4/9/d] *See also* **major sector**; **minor sector**; **radian**.

security Form of guarantee for the amount of a **loan**, usually asked for by a lender as an insurance against a **borrower** failing to meet the agreed terms of the loan. For example, a **mortgage** lender will hold the title deeds to a property so that, in the event of the lender failing to meet the terms of the mortgage, the lender can than take possession of the property and sell it to recover the amount of the mortgage outstanding. Alternative term: collateral.

■**segment** *1.* Part of a **circle** bounded by an **arc** and **chord**. *2.* Part of a line between two points. *3.* Part of a curve between two points on it. *See* Figure S2. [4/9/d]

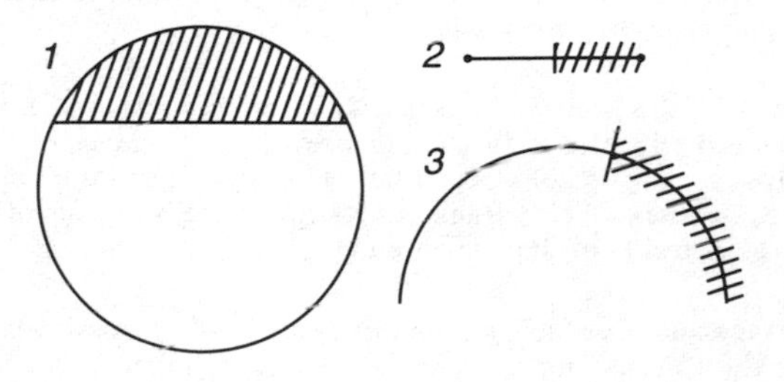

Fig. S2 *segment*

self-inverse Function $f(x)$ is self-inverse if $f \, . \, f(x) = x$ for all the x elements of its **domain**.

semi- Prefix meaning half, e.g. a semicircle is half a **circle** taken across the **diameter**.

sense *1.* Of lines of vectors, describing whether they are positive or negative according to which direction they point. They are positive in one direction, and negative in the opposite direction. The convention is often arbitrary. *2.* Of angles, describing whether they are positive or negative according to which **quadrant** they lie in.

septuagenarian Person reaching the ages of between 70 and 80.

■**sequence** Set of elements written in an ordered way. [3/6/b]

series Sum of an ordered **sequence** of terms. A series can be finite or infinite, converging or diverging. *See also* **arithmetic series**; **geometric progression**.

set (S) Collection of elements defined in a particular way, signified by the use of fretted brackets {...}. *Example*: {1p, 2p, 5p, 10p, 20p, 50p, 1·00} is the set of coins used in UK **currency**; {head, trunk, arms, legs} is the set of the most basic elements of the human body.

sexagesimal Describing a number system with the **base** 60, used for measuring: *1.* Time (60 seconds = 1 minute; 60 minutes = 1 hour); *2.* Angles in degrees of arc, where $1° = 60$ minutes (60′), 1 minute = 60 seconds (60″). Most **scientific calculators** will convert angles in sexagesimal notation to decimal notation, and vice-versa. *Example*: 25° 45′ 12″ = 25·75333° decimal. Sexagesimal was also used in the past by the Babylonians as the basis of angle measurement.

sf Abbreviation of **significant figure**.

■**shear** Sum of the vertical upward forces on a beam at all places. For a simple beam of negligible weight, fixed at one end, if a force F is applied downwards to the free end, the shear at any position on the beam will be equal to F. [4/10/c]

■**shearing transformation** **Transformation** that maps **parallel lines** onto parallel lines, as in Figure S3. One of the coordinate axes or coordinate planes remains fixed and the movement of any point is parallel to this line. **Area** is preserved with this transformation; shape is not. [4/10/d]

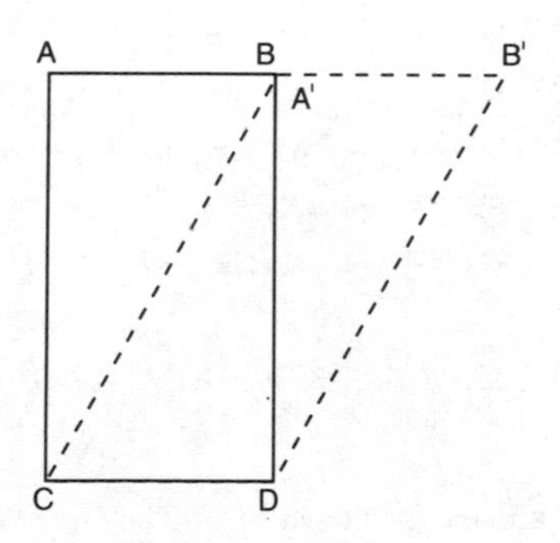

Fig. S3 *shearing transformation*

shm Abbreviation of **simple harmonic motion**.

■ **side** Flat surface bounding an object, e.g. a **cube** has six sides. [4/2/b]

sieve Process which finds **prime numbers**, discovered by the Greek mathematician Eratosthenes (lived *c.* 275–194 BC). He crossed out all the multiples of 2, 3, 5, 7 etc. from a list of **natural numbers**. All the numbers that remained were prime numbers. *Example*: to find all the prime numbers from 2 to 90:

	2	3	~~4~~	5	~~6~~	7	~~8~~	~~9~~	~~10~~
11	~~12~~	13	~~14~~	~~15~~	~~16~~	17	~~18~~	19	~~20~~
~~21~~	~~22~~	23	~~24~~	~~25~~	~~26~~	~~27~~	~~28~~	29	~~30~~
31	~~32~~	~~33~~	~~34~~	~~35~~	~~36~~	37	~~38~~	~~39~~	~~40~~
41	~~42~~	43	~~44~~	~~45~~	~~46~~	47	~~48~~	~~49~~	~~50~~
~~51~~	~~52~~	53	~~54~~	~~55~~	~~56~~	~~57~~	~~58~~	59	~~60~~
61	~~62~~	~~63~~	~~64~~	~~65~~	~~66~~	67	~~68~~	~~69~~	~~70~~
71	~~72~~	73	~~74~~	~~75~~	~~76~~	~~77~~	~~78~~	79	~~80~~
~~81~~	~~82~~	83	~~84~~	~~85~~	~~86~~	~~87~~	~~88~~	89	~~90~~

leaves

2, 3, 5, 7, 11, 13, 17, 19, 23, 29, 31, 37, 41, 43, 47, 53, 59, 61, 67, 71, 73, 79, 83, and 89.

sigma Greek name for the letter ‘s’. The Greek characters for both a small ‘s’ (σ) and a capital ‘S’ (Σ) are frequently used in mathematics. The capital Σ is used to represent **sum**; the small σ is used to symbolize the **standard deviation**. *Example*:

$$\sum_{p=2}^{p=6} p^3 = 2^3 + 3^3 + 4^3 + 5^3 + 6^3$$

$$\sigma^2 = \frac{1}{n} \Sigma \, (x - x^2)$$

sign Symbol used with an **operator** to denote whether a property is positive (+) or negative (−).

significance level Estimation or calculation of the probability of randomness, or deviation from the normal, in a statistical sample.

■**significant figure** (sf) Number which makes a contribution to a value, e.g. in 00·123, the numbers 1, 2, 3 are significant. Significant figures are used to denote the accuracy of an answer, e.g. the number 43·98 is correct to two decimal places but it is correct to 4 significant figures. [2/5/g] *See also* **rounding off**.

■**similar angle** Angle which is equal to another. [4/5/b]

similar figure Figure which has a **corresponding angle** equal to that in another figure. Corresponding sides are also in proportion between the two figures concerned.

similar triangles Triangles that have equal corresponding angles and corresponding sides in proportion to each other. The proof lies in applying the theorem of **Pythagoras** to the triangles in question. *Example*: take two equiangular triangles ABC and AEF as in Figure S4.

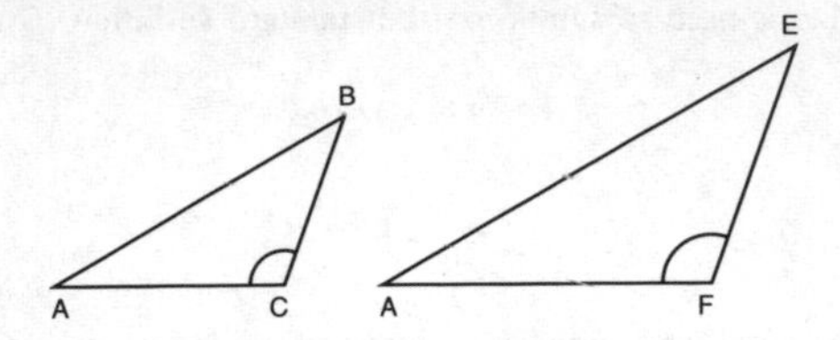

Fig. S4 *similar triangles (1)*

Now superimpose them and construct right angles from the extended line AF to the **vertices** B and E, calling the points where the right angles lie on extended AF D and G respectively, as in Figure S5.

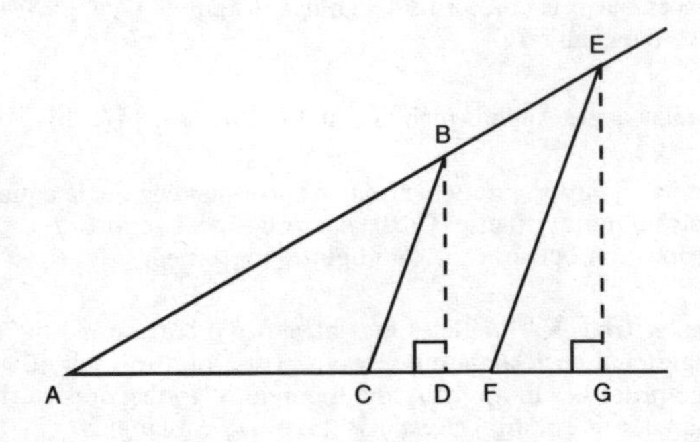

Fig. S5 *similar triangles (2)*

Taking the triangle ABD, it can be shown from Pythagoras' theorem that

$$BD^2 + CD^2 = BC^2$$

$$BD^2 + AD^2 = AB^2$$

Therefore, as BD is to CD, so BD is to AD. From this it follows that

$$\frac{CD}{AD} = \frac{BC}{AB}$$

Again, using Pythagoras' theorem, we can show for triangle AEG that

$$EG^2 + FG^2 = EF^2$$

$$EG^2 + AG^2 = AE^2$$

Therefore, as EG is to FG, so EG is to AG. From this it follows that

$$\frac{FG}{AG} = \frac{EF}{AE}$$

Taking the conclusions from both triangles, it follows that

$$\frac{BC}{EF} = \frac{AB}{AE} = \frac{AC}{AF}$$

Since corresponding sides in similar triangles are proportional to each other then, whatever the lengths of the sides of the triangles, the ratios of the sides to each other are constant. These constant ratios are the basis of **trigonometric ratios** such as **cosine**, **sine** and **tangent**.

simple closed curve Curve which is closed and does not cut over itself.

simple harmonic motion (shm) Type of motion associated with pendulums or springs. It is defined for a moving object in such a way that the acceleration is proportional to the distance from a fixed point in the line of motion and is directed towards that point. It is often referred to as an oscillation.

simple interest **Interest** at an agreed rate (R) charged to a **borrower** or paid to an investor, calculated on the whole amount of the sum lent or invested (the principal, P) over the whole period of the loan or investment (time, T), usually expressed as a percentage. *Example*: find the simple interest on £600·00 at 12% over 5 years. P = principal; R = rate; T = time in years; I = interest, and $I = \frac{PRT}{100}$:

$$I = \frac{(600 \times 12 \times 5)}{100} = £360{\cdot}00$$

See also **compound interest**.

■ **simplify** In mathematics, to perform one operation on an expression or number of expressions (such as finding the **lowest common multiple** of a number of fractions) to make it easier to perform another operation, such as multiplication or division. In algebraic expressions simplifying is often achieved by cancelling or factorizing. [3/10/b] *See* **cancellation**; **factorize**.

■ **Simpson's rule** Rule for finding the approximate area under a curve. [3/10/c] *Example*: using Figure S6, $y = ax^2 + bx + c$; a,

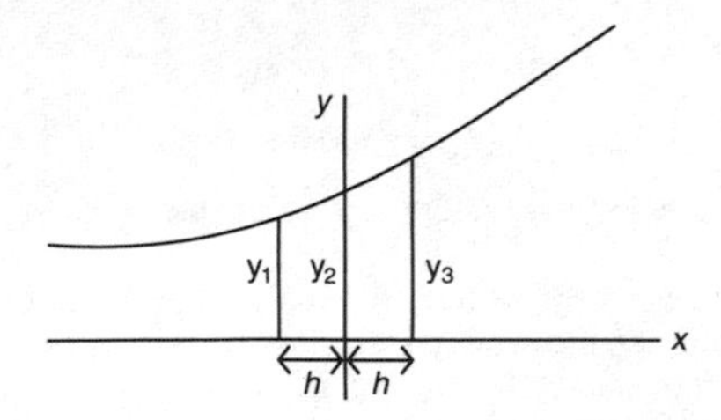

Fig. S6 *Simpson's rule*

b and c can be found from:

$$y_1 = a(-h)^2 + b(-h) + c$$
$$y_2 = 0 + 0 + c$$
$$y_3 = ah^2 + bh + c$$
$$y_1 + y_3 = 2ah^2 + 2c$$

Area under a quadratic arc is:

$$\mathrm{A} = \int_{-h}^{h} (ax^2 + bx + c) \,.\, dx$$
$$= [\tfrac{1}{3} \,.\, ax^3 + \tfrac{1}{2} \,.\, bx^2 + cx]^{h}_{-h}$$
$$= \tfrac{2}{3} \,.\, ah^3 + 2ch$$

This can be written as

$$\mathrm{A} = \tfrac{1}{3} \,.\, h(2ah^2 + 6c)$$

but

$$2ah^2 + 6c = y_1 + y_3 + 4y_2$$

$$\therefore \mathrm{A} = \tfrac{1}{3} . h(y_1 + 4y_2 + y_3)$$

See also **mid-ordinate rule**; **trapezoidal rule**.

■**simultaneous equations** Set of two algebraic equations that are all true for the same particular values of their **variables**. [3/7/g] *Example*: solve the equations

$$\textbf{(1)} \quad 5x + 2y = 16$$

and

$$\textbf{(2)} \quad 3x + 4y = 18$$

First, multiply (**1**) by 3 and (**2**) by 5, which gives

$$\textbf{(3)} \quad 15x + 6y = 48$$

$$\textbf{(4)} \quad 15x + 20y = 90$$

Now, subtract (**3**) from (**4**)

$$\begin{array}{r} 15x + 20y = 90 \\ -15x + 6y = 48 \\ \hline 14y = 42 \\ y = 3 \end{array}$$

Now substitute $y = 3$ in (**1**), which gives

$$5x + 2(3) = 16$$

$$5x + 6 = 16$$

$$5x = 10$$

$$x = 2$$

Therefore the solution to the two original equations is $x=2$, $y=3$. *See also* **elimination**.

sin Abbreviation of **sine**.

sine (sin) **Trigonometrical ratio**. The sine of angle θ in a right-angled triangle is equal to the **ratio** of the length of the side opposite the angle to the **hypotenuse** of the triangle. *Example*: in Figure S7 the right-angled triangle ABC is shown, with sides a, b and c and $\angle$B marked as θ to show that its size is unknown. The two sides of the triangle relevant to the sine are drawn thicker than the third side.

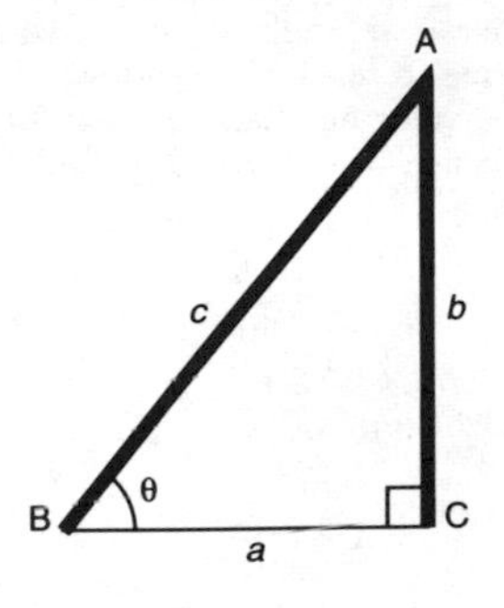

Fig. S7 *sine*

From the properties of **similar triangles** it is known that the ratio AC : AB is a constant; it is this ratio which is meant by the term sine. Using the terms a, b and c it can also be said that

$$\sin \theta = \frac{b}{c}$$

and

$$b = c \sin \theta$$

$$c = \frac{b}{\sin \theta}$$

Provided that two of the variables are known in a sine problem, the third can always be found. *Example*: a builder wants to repair a window 6 m from ground level, and positions a ladder at an angle of 70°; to what length will the ladder be extended to reach the window? Using the triangle in Figure S7, the value 6 m can be given for b, and 70° for the angle θ. The side which has to be found, therefore, is c. Using the formula

$$c = \frac{b}{\sin \theta}$$

add the values, which gives

$$c = \frac{6}{\sin 70°}$$

First, find sin 70° = 0·93969; now divide 6 by the sine = 6·38. The ladder, therefore, will have to be extended to a length of 6·38 m. The value of a sine can be found either by consulting a table of natural sines, or using a calculator with a 'sin' key. With a calculator, enter the number of degrees, and then press the 'sin' key. Tables of natural sines list the sines in the way shown below

	0′	6′	12′	18′	24′
41°	·6561	6574	6587	6600	6613
42	·6691	6704	6717	6730	6743
43	·6820	6833	6845	6858	6871

The value in degrees is given at the extreme left of the table. A full table will give sines for 0–54 minutes (′). Sine tables can also be used in reverse to find the angle when the sine is known: look for the sine value on the table, look at the top of the table to check how many minutes of arc it equals, and then to the extreme left to find the angle. Sine tables can also be used to find the angle of a **cosine**, because in a right-angled triangle, $\angle$A and $\angle$B are **complementary**. Therefore sin A = cos B, and $\cos \theta = \sin (90° - \theta)$.

■ **sine rule** Mathematical rule which states that for a triangle ABC of sides *a*, *b*, and *c* (as in Figure S8):

$$\frac{\sin A}{a} = \frac{\sin B}{b} = \frac{\sin C}{c}$$

OR

$$\frac{b}{c} = \frac{\sin B}{\sin C}$$

$$\frac{a}{b} = \frac{\sin A}{\sin B}$$

$$\frac{a}{c} = \frac{\sin A}{\sin C}$$

The rule can be used for all triangles where one side and two angles are known, or where two sides together with the opposite angle are known. *Example*: in a triangle ABC,

$\angle A = 40°$, $a = 14$, and $b = 7$. Find $\angle B$. Since a and b are known, the formula

$$\frac{a}{b} = \frac{\sin A}{\sin B}$$

can be used. Inserting the values supplied gives

$$\frac{14}{7} = \frac{\sin 40°}{\sin B}$$

$$\frac{2}{1} = \frac{0{\cdot}64278}{\sin B}$$

$$\sin B = \frac{0{\cdot}64278}{2}$$

$$= 0{\cdot}32313$$

$$= 18{\cdot}7$$

$$= 18°\ 44'$$

To prove the solution is correct, use the formula

$$\frac{\sin A}{a} = \frac{\sin B}{b}$$

Inserting the values supplied gives

$$\frac{0{\cdot}64278}{14} = \frac{0{\cdot}32313}{7}$$

which is correct. Since two of the angles of the triangle are known, the third can be found:

$$180° - (40° + 18°\ 44') = 121°\ 16' = 121{\cdot}3$$

Now that the third angle is known, the length of c can be

found using the formula

$$\frac{\sin B}{b} = \frac{\sin C}{c}$$

Again, inserting the values supplied gives

$$\frac{0{\cdot}32313}{7} = \frac{\sin 121{\cdot}3}{c}$$

$$= \frac{0{\cdot}85445}{c}$$

Therefore

$$c \times 0{\cdot}32313 = 7 \times 0{\cdot}85445$$

$$= \frac{5{\cdot}98121}{0{\cdot}32313}$$

$$= 18{\cdot}510$$

There is also an extension to the sine rule which states that:

$$\frac{\sin A}{a} = \frac{\sin B}{b} = \frac{\sin C}{c} = 2R$$

where R is the radius of the **circumscribed circle** to the triangle. [4/10/b]

singular Denoting: *1.* unity; *2.* one of; *3.* of order one.

singular matrix Matrix where the **determinant** is equal to zero.

sinh Abbreviation for the **hyperbolic function** $\frac{1}{2}(e^x - e^{-x})$.

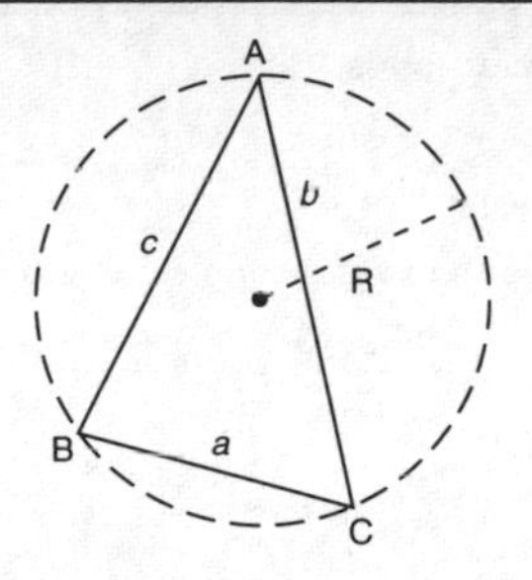

Fig. S8 *sine rule*

SI units System of measurement based on the metre, kilometre and second (m.k.s.), now generally established as the standard system of measurement in scientific calculation. SI is an abbreviation of Système Internationale d'Unités. There are seven basic units, as follows:

unit	*symbol*	*quantity*
metre	m	length
kilogram	kg	weight
second	s	time
kelvin	K	temperature
ampere	A	electric current
mole	mol	amount of substance
candela	cd	luminous intensity

■**skew distribution** Non-symmetrical **distribution** of data on a graph, as shown in Figure S9. [5/10/b]

skew lines Straight lines which do not intersect and are not parallel to each other.

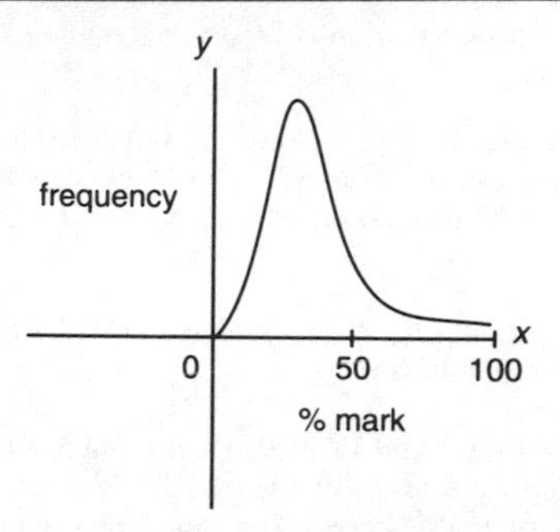

Fig. S9 *skew distribution*

slant height Perpendicular distance from a point X on an inclined plane to the horizontal.

slide rule Manual device for performing calculations consisting of two parallel horizontal logarithmic scales, one set into the other so that the two can be slid past each other to allow multiplication and division by adding and subtracting numbers. Slide rules can be used to find numbers to given powers and roots, **trigonometric ratios** and **reciprocals**. Until the advent of the electronic **calculator**, the slide rule was very widely used, particularly for calculations in scientific and engineering mathematics. Different versions of the device were in use from as early as the 17th century (i.e. not long after Napier's invention of **Napierian logarithms**), but the form in which it became widely used was evolved during the 19th century. *See also* **logarithm**.

small circle Section taken through a **sphere** which does not pass through its centre, for example the lines of **latitude** used to help determine position on the surface of the Earth. *See also* **longitude**.

solar day Period of time it takes the Earth to make one complete rotation on its **axis**. This can vary from between 24 hours 0 minutes 30 seconds and 23 hours 59 minutes 39 seconds.

solid Object having the three dimensions of length, breadth and height; a **polyhedron**.

solid of revolution Solid formed when a plane surface is revolved about a line, as in Figure S10. If a curve is revolved about the *x* or *y* axis between certain limits then the area or volume of the surface formed can be obtained by using integral calculus (*see* **integration**).

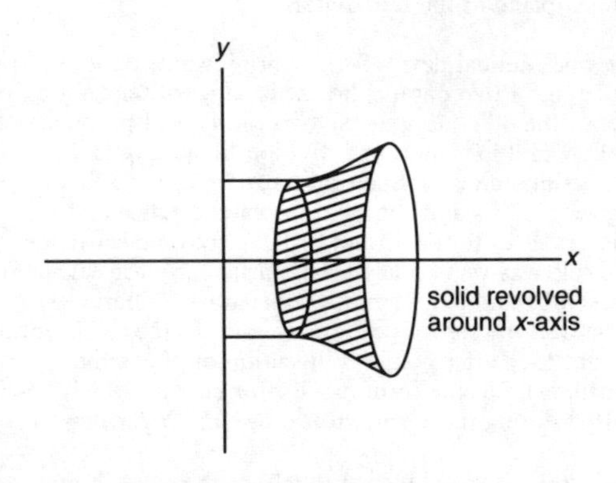

Fig. S10 *solid of revolution*

solution Resolution or answer to a mathematical question or problem. *See also* **satisfy**.

■**speed** Rate of change of displacement with time. It is a **scalar** since no **direction** is specified, and has units of length divided by time, e.g. metres per second (m/s), kilometres per hour (km/h), and miles per hour (mph). [2/6/h]

■**sphere Solid** obtained by rotating a semicircle about its **diameter**. The surface area of a sphere can be calculated with the formula $4\pi r^2$, where r is the radius of the sphere; for calculating the volume of a sphere, the formula $\frac{4}{3}\pi r^3$ can be used. [4/2/a]

spiral Curve produced by moving a point B about another point C such that the distance r from point C has a fixed relation to the angle of rotation A; r is the **radius vector** and A is the vectorial angle.

sq Abbreviation for **square**, sense *2.*

■**square** *1.* Four-sided plane figure with all its sides equal and all angles right angles. The area of a square of side l is l^2. *2.* (sq) Number obtained when another number is multiplied by itself. It is indicated by the index (power) 2; e.g. $4^2=16$. [4/2/a]

■**square matrix** Matrix which contains the same number of columns as rows. [4/10/d]

■**square root** Of a number, another number that when multiplied by itself gives the original number. It is indicated by the symbol $\sqrt{}$ or the index (power) $\frac{1}{2}$; e.g. $\sqrt{16}$ (or $16^{\frac{1}{2}}$) = 4. [3/5/c] *See also* **root**.

sr Abbreviation for **steradian**.

■**standard deviation** (sd) **Root** of the **mean** of the squares of the differences from their mean of a number of observations. It is denoted by the symbol σ and calculated from the formula:

$$\sigma^2 = \frac{1}{n} \Sigma(x - \bar{x})^2$$

where n is the number of observations, and σ^2 is the sum of the squares of the n deviations. [5/10/a] *Example*: in a group of observations with values 3, 7, 8, and 4, the mean of the number of observations is

$$\frac{3+7+8+4}{4} = 5{\cdot}5$$

The deviations from the mean are

$$2{\cdot}5,\ 1{\cdot}5,\ 2{\cdot}5,\ 1{\cdot}5$$

The squares of the deviations are

$$6{\cdot}25,\ 2{\cdot}25,\ 6{\cdot}25,\ 2{\cdot}25$$

The mean of the squares is

$$\frac{6{\cdot}25+2{\cdot}25+6{\cdot}25+2{\cdot}25}{4} = \frac{17}{4}$$

$$= 4{\cdot}25$$

The standard deviation is

$$\sqrt{4{\cdot}25} = 2{\cdot}06155$$

Standard deviation can also be found using a **scientific calculator** with statistical function keys (*see* Appendix II).

■**standard form** Simple form in which an **equation** can be written. Any equation can be transformed into its appropriate standard form. [5/10/a] *Example*: all cubic equations can be of the form:

$$x^3 + ax + b = 0$$

or all binomial equations can be written as

$$x^2 + 2xy + y^2$$

■**standard index form** Expression of all numbers in the form $N \times 10^n$ where N is a number between 1 and 9 and n is an **integer**. [2/8/b] *Example*: 53,400 can be written in standard index form as 5.34×10^4. Values in standard index form can be found on calculators with a 'SCI' mode. Alternative term: **scientific notation**.

■**stationary point** Point on a graph where the **tangent** to the curve is parallel to the x axis (see Figure S11). All turning points are stationary points but the opposite is not always true. A maximum value of the function $f(x)$ happens when

$$\frac{dy}{dx} = 0$$

and

$$\frac{d^2y}{dx^2} < 0$$

i.e., is negative. A minimum occurs when

$$\frac{dy}{dx} = 0$$

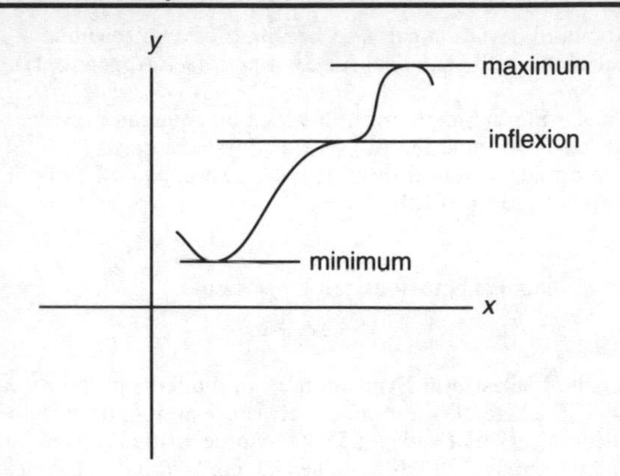

Fig. S11 *stationary point*

and

$$\frac{d^2y}{dx^2} > 0$$

i.e., is positive. Note that if

$$\frac{d^2y}{dx^2} = 0$$

then this is a special case called a point of **inflexion**. [3/9/f]

statistical analysis Evaluation of data by the use of **statistics**.

statistical mechanics Branch of **statistics** concerned with the study of the properties and behaviour of the component particles of a macroscopic system.

statistics Branch of **science** concerned with the collection and classification of numerical data and facts, and their interpretation in mathematical terms, especially the determination of **probabilities**.

steradian (sr) **SI unit** of solid angle which is equal to the angle at the centre of a **sphere** subtended by the part of the surface whose area is equal to the square of the **radius**. *See also* **radian**.

■**straight line (equation of)** In **coordinate geometry**, the equation of a straight line in standard form is $y = mx + c$ where m is the gradient of the line and c is the intercept point on the y axis (see Figure S12). [3/7/h] Alternative term: **linear equation**.

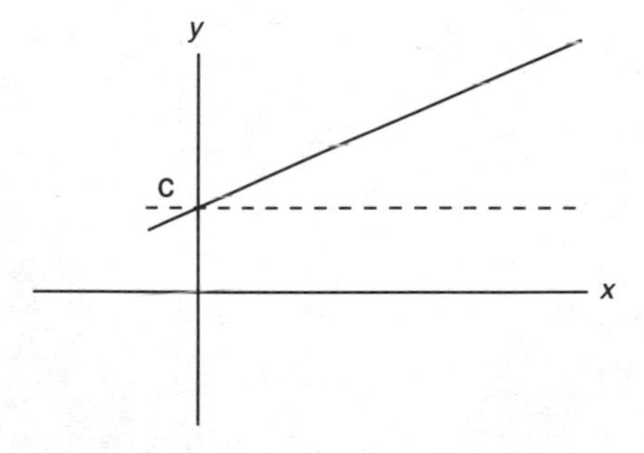

Fig. S12 *straight line*

subgroup Subset of a group.

subset Set within a set, which can be represented as a **Venn diagram**, symbol $\subset$.

■**substitution** Putting numerical values in place of the variables in an algebraic expression to arrive at a solution of the expression. [2/8/c] *Example*: if $a=2$ and $b=3$ what is $3a+b^2$? Substituting the numerical values supplied gives: $3\times 2+3^2=6+9=15$. [2/8/c] *See also* **elimination**.

subtend If two points X and Y on a curve are joined to a point P then XY is said to subtend the angle XPY at P, as shown in Figure S13. *See* **radian**.

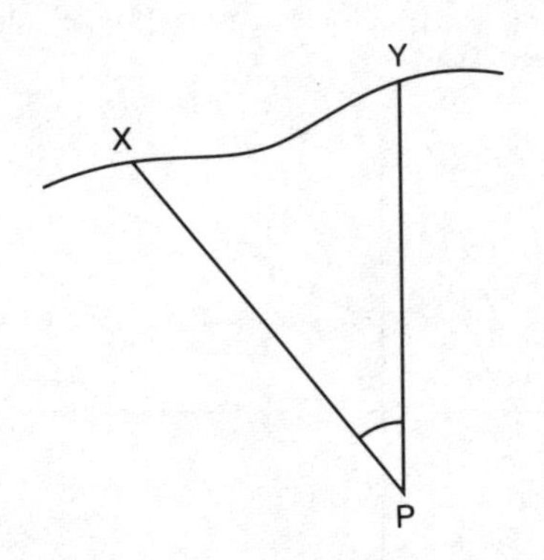

Fig. S13 *subtend*

subtraction Binary operation to find the remainder when part of the whole is removed. It is the opposite of addition.

subtraction of fractions If the **denominators** in two fractions are different where one is being subtracted from another, they have to be made similar. Then the numerators can be subtracted leaving a single fractional answer. *Example*:

$$\frac{6}{16}-\frac{4}{32}=\frac{6}{16}-\frac{2}{16}=\frac{4}{16}=\frac{1}{4}$$

sufficient Adequate number or amount to complete a particular task.

sum (S) Total, the outcome of **addition.**

super computer Very fast, complex computer with a very large memory capacity.

■ **supplementary angle** Angle which, when added to another angle, gives a sum of 180°, i.e. a straight line. [4/5/b] *Example*: the supplementary angle of 32° = 148°.

■ **surd** Irrational number, for example the square root of 2 ($\sqrt{2}$) which, to 11 decimal places = 1·41421356238. The precise value (like that of π) cannot be finally calculated. [2/9/a]

■ **symmetry** Property of being symmetrical, i.e. having precisely the same shapes on each side of or around a point, axis or plane. [4/3/b] [4/4/e] *See also* **axis of symmetry**.

synchronisation Fixing of different events to take place at the same point in time. For example, four people make a journey to the same spot from the same beginning point, but each following a different route to their common destination.

If the four people want to measure how long it takes each of them to follow their respective routes, they first need to make sure that their respective watches all tell exactly the same time when they all begin their journeys. All four people therefore synchronise their watches at the beginning of their journeys to make sure that their starting points in time are identical. Almost all machines are built with synchronised parts. *See also* **leap year**.

T

T Number 10 in the **duodecimal** number system.

T Symbol for **tera-**, $\times 10^{12}$.

t Symbol for time. *See* **equations of motion**.

tan Abbreviation for the **trigonometrical ratio tangent**.

■ **tangent** In geometry, *1.* a line that touches a curve at one point without cutting it. At the point of contact, the tangent has the same slope as the curve. [3/9/f] *2.* (tan) **Trigonometrical ratio**. For an angle θ in a right-angled triangle, the tangent is the ratio of the length of the opposite side to the adjacent side. [4/8/b] [4/9/a] Tangents are usually signified by the use of the abbreviation 'tan', and the size of the angle when it is not known by the Greek letter θ. *Example*: in trigonometrical problems, it is usual to label the angles in the triangle according to which **vertex** they occupy. So, in triangle ABC in Figure T1,

$$\tan \mathrm{B} \text{ (or } \tan \theta) = \frac{\mathrm{AC}}{\mathrm{BC}} = \frac{b}{a}$$

where $a=\mathrm{BC}$, $b=\mathrm{AC}$ and $c=\mathrm{AB}$. From this it follows that

$$a \times \tan \mathrm{B} = b$$

and

$$a = \frac{b}{\tan \mathrm{B}}$$

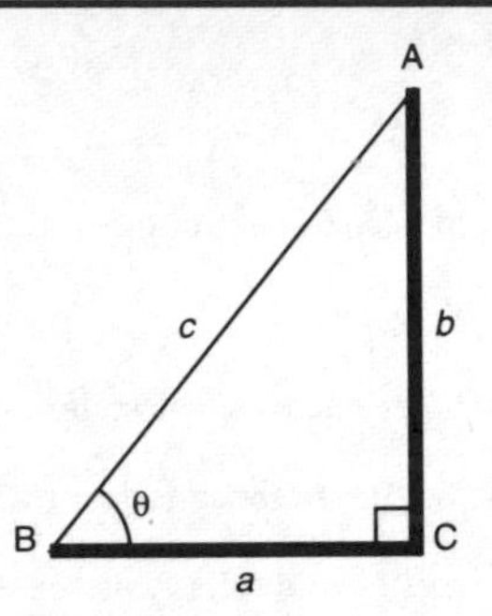

Fig. T1 *tangent* (*sense 2.*)

As long as two of the values are known, it is therefore possible to find the third. *Example*: a person in front of a house 20 m high measures the shadow of the house cast by the Sun as 6 m long. At what angle is the Sun? The height of the house can be taken as side AC in the triangle in Figure T1, and the length of the shadow as side BC. Applying the rule above, the size of $\angle$B can be found using $\tan B = \frac{b}{a}$, therefore

$$\tan \theta = \tfrac{20}{6} = 3{\cdot}3333' = \tan 73{\cdot}30 = 73^\circ\ 18'$$

Another problem: if the eyes of a person 1·7 m tall standing at the bottom of a staircase are raised at an angle of 59° 32′ to meet the eyes of another 1·7 m tall person at the top of the staircase, how deep is the staircase? Using Figure T1 again, it is apparent that it is *a* that we need to find.

Therefore

$$a = \frac{1 \cdot 7}{\tan 59^\circ\ 32'}$$

$$\tan 59^\circ\ 32' = 1 \cdot 6996$$

Therefore

$$a = \frac{1 \cdot 7}{1 \cdot 6996} = 0 \cdot 99982 \text{ m} = \text{approx 1 m.}$$

When measuring angles, it is possible to describe them both in terms of degrees and minutes (59° 32′) or as a decimal (59·5344). Also, as the size of an angle increases, so too does the tangent of the angle: tan 35° = 0·70020, tan 45° = 1, tan 60° = 1·73205, and tan 5·67128. This is because, as the angle increases, the sides of the triangle are further and further apart, and so their ratios increase. Tangents of angles can readily be found either in tables of natural tangents, or using a calculator which has a 'tan' key. On tables, the information is presented as below

	0′	6′	12′	18′
46°	1·0355	0392	0428	0464
47	1·0724	0761	0799	0837
48	1·1106	1145	1184	1224

The angle is shown in degrees at the extreme left hand of the rows, and the tangents are then given for intervals of 6 minutes up to 54′. Note that the **characteristic** and **mantissa** are given only in the second (0′) column; the mantissa alone appears in the other columns. The tables can also be used in reverse for finding the size of an angle if the tan is known by finding the relevant tangent value on the table, and then reading from right to left to the degrees column. If the tan 1·0799 has been found in a calculation, look for the mantissa 0799 on the table, look at the top of the column to check how many minutes it equals, and then read to the left to find the degrees = 47° 12′.

tanh Abbreviation for the **hyperbolic function**

$$\frac{\sinh x}{\cosh x}$$

tape deck Unit that records audio or video signals onto, or plays back signals from, magnetic tape in cassettes or on reels.

tape punch Machine that produces punched paper tape, either operated by a **keyboard** as part of a computer **input device** or automatically as an **output device** driven by a computer.

tape reader Input device that feeds **data** off punched paper tape into a computer.

Taylor series Mathematical expression that represents a **function** as the sum of an infinite **series** of terms. It was named after the British mathematician Brook Taylor (1685–1731), and is usually given as:

$$f(x) = f(a) + f'(a)(x-a) + f''(a) \cdot \frac{(x-a)^2}{2!} + f'''\frac{(a)(x-a)^3}{3!}$$

$$+ \ldots f^{(n-1)} \frac{(a) \cdot (x-a)^{n-1}}{(n-1)!} + R_n$$

See also Appendix III.

tera- (T) Prefix in the **metric system** signifying $\times 10^{12}$.

terminal decimal Decimal quantity that has a finite number of digits after the decimal point, for example 0·75. *See also* **recurring decimal**.

ternary Describing a number system that has the base 3.

tessellation Patterns using congruent **equilateral triangles** or hexagons or squares. Typically, patterns made by arrays of similar tiles form tessellation.

tetrahedron Polyhedron with four **faces**, all of which are **equilateral triangles**. Tertrahedra are one of the five regular polyhedra. *See* **polyhedron**.

theorem General conclusion in science or mathematics which makes certain assumptions in order to explain observations. *See also* **hypothesis**.

tiling *See* **tessellation**.

■ **tonne** Unit measure of mass on the **S.I. system**; 1 tonne = 1,000 kg. Not to be confused with the **British unit** Ton = 2,240 pounds = 1,016 kg, or with the US ton = 2,000 lb = 907·2 kg. [2/5/h]

topology Branch of **geometry** concerned with the properties of surfaces, without regard to absolute size or shape.

torque Turning **moment** produced about an **axis** by force acting at right angles to a **radius** from the axis.

torus Solid figure that resembles a tyre or doughnut. If the radius of the solid part is r and the radius of the whole ring is R, its area is $4\pi^2rR$ and its volume is $2\pi^2r^2R$. Alternative name: anchor ring.

■**total probability** Probability of all possible things happening, i.e. certainty. For example, the probability of success plus the probability of failure, which equals 1. [5/4/g,h,i]

trajectory Path followed by a moving **projectile** acted upon by gravity or other forces.

transcendental *1.* **Irrational number** that is not the **root** of a **polynomial** equation. *2.* Function that is not a finite **polynomial** equation (e.g. **logarithmic** or **exponential function**).

■**transformation** *1.* Process for obtaining from a mathematical point, line etc. a corresponding point or line. *2.* Process for changing the subject in a mathematical formula. [3/8/a]

transitive Relationship which, if it applies from A to B and from B to C, also then applies from A to C. *Example*: if $x = y$ and $y = z$, then $x = z$.

translation Movement by the same amount, and in the same direction, by every point in a line or in a plane figure.

■**transpose** In mathematics, to change a term from one side of an equation to another, often referred to as changing the subject of a formula. A **matrix** can also be transposed if rows and columns are interchanged. [2/8/c] [3/4/d]

transversal Line that intersects two or more other lines, thus creating **alternate angles**.

trapezium Four-sided figure (quadrilateral) with two parallel sides. If these sides have lengths p and q, and they are h apart, the area of the trapezium is $\frac{1}{2}(p+q)h$.

trapezoid Quadrilateral that has none of its sides equal to any of the other three, and neither of the two pairs of sides are parallel.

■ **trapezoidal rule** Rule for finding the area under a curve. The area is divided into vertical strips of equal width. The total area under the curve is equal to the width of the strips multiplied by half the sum of the first and last ordinates plus the sum of the remaining ordinates. [3/10/c] *See also* **mid-ordinate rule**; **Simpson's rule**.

tree diagram *See* **decision tree diagram**.

■ **trial** Process by means of which the properties of quantities are tested, often by experiment, e.g. when collecting statistical data. [2/5/f]

■ **trial and improvement method** Series of estimates used to find the solution to a problem. The final answer is arrived at after each successive **approximation** or estimate yields a slightly improved result. Finally a close approximation to the required answer is achieved. [2/5/f] *Example*: to find the value of π, first take a circular object; wrap a piece of string around its circumference; measure the length of string = 290 mm. Next, measure the diameter of the same object with a ruler = 89 mm. Next divide 290 by 89 = $3\frac{23}{89}$. Express this as a decimal = 3·25842. Subtract 2 mm from diameter and circumference measurements to allow for error, divide as before = 3·31034. Add 2 mm to allow for error; divide as before = 3·20879. Add the three results and find the mean = 3·25918.

■**triangle** Three-sided **polygon** in which the sum of the **interior angles** equals 180°. Scalene triangles have all three sides unequal in length; isosceles triangles have two sides equal in length, and equilateral triangles have all three sides equal in length. The area A of any triangle is equal to half the product of the length of the base b and the altitude h (perpendicular distance from the apex to the base); i.e. $A=\frac{1}{2}bh$. The triangle is one of the most basic geometric forms, and has been used to discover and prove many mathematical **axioms**, such as the theorem of **Pythagoras**, and the **trigonometric ratios**. [4/2/a] *See also* **Hero's formula**.

■**triangle of vectors** Triangle whose sides represent the magnitude and direction of three **vectors** in **equilibrium** acting at a point, as in Figure T2, where $\vec{A}+\vec{B}=\vec{C}$. [4/9/e] *See also* **parallelogram of vectors**.

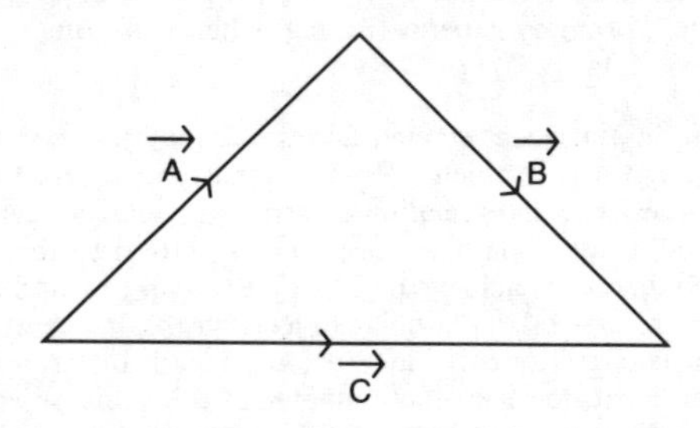

Fig. T2 *triangle of vectors*

■**triangle of velocities** Triangle of **vectors**, in which each vector denotes a **velocity**. [4/9/e]

trigonometrical function Alternative name for **trigonometrical ratio**.

trigonometrical ratio Mathematical relationship between the angles and sides of a right-angled triangle. **Sine**, **cosine**, and **tangent** are the most commonly used of these relationships. They are all constant **ratios**; i.e. whatever the size of an angle in a triangle, and whatever the lengths of the sides, the ratios of given sides to other sides is always the same. The constancy of these ratios allows solutions to be found for almost any triangle where the length of a side or the size of an angle is unknown. The basic figure for trigonometric ratios is shown using **Cartesian coordinates** in Figure T3, where x and y are the axes, (x, y) are the **ordinate** and **abscissa** respectively, and θ is the angle. The principal ratios

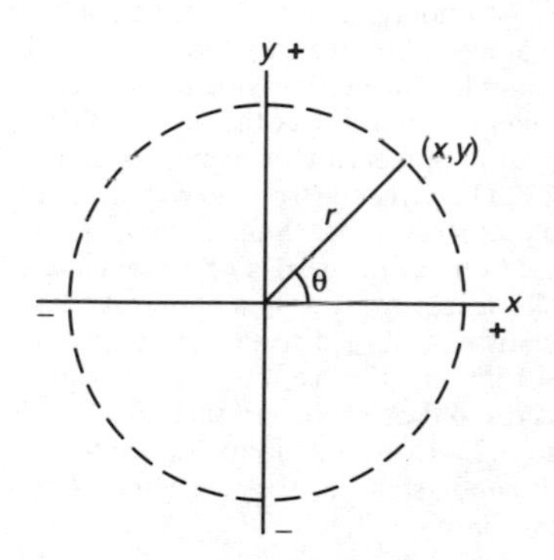

Fig. T3 *trigonometrical ratio*

are

$$\sin \theta = y/r$$

$$\cos \theta = x/r$$

$$\tan \theta = \sin \theta / \cos \theta = y/x$$

Alternative name: **trigonometrical function**. *See also* **cosecant**, **cotangent** and **secant**.

■ **trigonometry** Branch of mathematics based on **trigonometrical ratios**, i.e. the relationships between the angles and sides of **triangles**. In its modern form, trigonometry is a highly-evolved area of mathematics, and is used in fields such as engineering, astronomy, architecture and surveying. The roots of the subject, however, lie thousands of years in the past. It is possible that the Babylonians understood some of the basic principles, but it was the ancient Greeks who began a systematic development of the concepts underlying modern trigonometry. The Greek astronomer Ptolemy constructed the first table of **sines** in the 2nd. century AD using a theorem based on the properties of a **cyclic quadrilateral**. Over the following centuries, Hindu and Arab mathematicians contributed to the development of the theory until, in the 16th and 17th centuries European mathematicians and astronomers such as Nicholas Copernicus (1473–1543) and John Napier fully developed what we now understand as trigonometry. [4/9/a]

trinomial Polynomial with only three terms.

truncated Describing a solid figure formed by cutting through a **cone** or **prism** with two planes. If the planes are parallel, the figure is a **frustum**.

■ **turning point** *See* **stationary point**. [3/9/f]

U

union *1.* Summation of two **sets**, symbol $\cup$. *Example*: the union of the two sets {h, i, k, m} and {j, l} is {h, i, j, k, l, m}. *2.* Set containing common elements of two other sets. *Example*: of two sets {potatoes, carrots, celery, onions} and {tomatoes, leeks, potatoes, parsnips, onions}, the union is {potatoes, onions}.

■**unit** *1.* Defined constant quantity used for measurement on an appropriate **scale**. *Example*: inches and millimetres are units on the **f.p.s. system** and **metric system** respectively. To find the equivalent of a quantity using one set of units in another set of units, for example to change inches to millimetres, it is necessary to use a **conversion** factor. Some calculators have these included, and will perform conversions automatically. [2/2/h] *2.* The numbers 1–9 are units in the **denary** system. *See* **abacus**; **long division**; **long multiplication**.

■**unit matrix** Square, diagonal matrix in which the diagonal elements are equal to 1, i.e.:

$$\begin{pmatrix} 1 & 0 \\ 0 & 1 \end{pmatrix}$$

[4/10/d]

universal set **Set** containing all elements.

unknown Value that is found by means of a mathematical operation. Unknowns are often represented by terms such as x and y. *Example*: in $3x^2+6x-3=0$, x is the unknown.

upper bound *1.* Maximum level of any mathematical function, e.g. the upper bound of the **sine** of an angle is 1. *2.* Number of the largest member of a **set**.

V

V *1.* Abbreviation for **volume**. *2.* Number 5 in the **Roman numeral** system.

v Symbol for **vector**.

v *1.* Symbol for **frequency**. *2.* Symbol for **velocity**.

value Result or answer of an expression after some mathematical operation has taken place. Usually denoted by a number.

value added tax (VAT) Point-of-sale tax levied on particular goods. In 1992, UK VAT stood at 17·5%. Therefore an article priced at £899·95 subject to VAT would be sold at

$$\pounds 899{\cdot}95+\left(\pounds 899{\cdot}95\times\frac{17{\cdot}5}{100}\right)=\pounds 899{\cdot}95+\pounds 157{\cdot}49$$

$$=\pounds 1057{\cdot}44$$

Using a calculator with parentheses (brackets) keys, key 899·95, then +, then open bracket, then 899·95, then ×, then 17·5, then INV, then %, then close bracket, then =. Without parentheses keys, key 899·95, then ×, then 17·5, then INV, then %, then =, then +, then 899·95, then =.

variable *1.* In mathematics, symbol such as x or y that can have any number of different values. *Compare* **constant**; *see also* **unknown** *2.* In computing, block of data that is stored at different locations during the operation of a program. *3.* In astronomy, alternative name for **variable star**.

variance For a set of numbers, the mean of the squares of the **deviations** of each number from the arithmetic **mean** of the set. Its square root is the **standard deviation**. *Example*: in the set

$$\{6, 7, 8, 9\}$$

the arithmetic mean is

$$\frac{6+7+8+9}{4} = 7{\cdot}5$$

the squares of the deviations are 2·25, 0·25, 0·25, 2·25. The mean of the squares (the variance) is

$$\frac{2{\cdot}25+0{\cdot}25+0{\cdot}25+2{\cdot}25}{4} = 1{\cdot}25$$

VAT Abbreviation for **value added tax**.

VDU Abbreviation of **visual display unit**.

■**vector** In mathematics, a quantity with both direction and magnitude (a quantity with magnitude only is a **scalar**), usually symbolized by a letter with an arrow above it (e.g. $\vec{A}$), or by **v**. [4/8/d] *Example*: the speed of a ship travelling at 12 **knots** is a scalar quantity; the speed of a ship travelling at 12 knots in the direction South-West is a vector quantity.

■**velocity** (v) Rate of movement in a particular direction; distance travelled per unit of time. It is a **vector** quantity, unlike **speed** (which is a **scalar** quantity, for which direction is not specified). [2/6/h] *See* **equations of motion**.

Venn diagram Graphical representation using rectangles and circles, of operations or relationships with **sets**. *Example*: a rectangle, as shown in in Figure V1, represents the **universal set**, often denoted by $\in$. In Figure V2, A equals **subset** A,

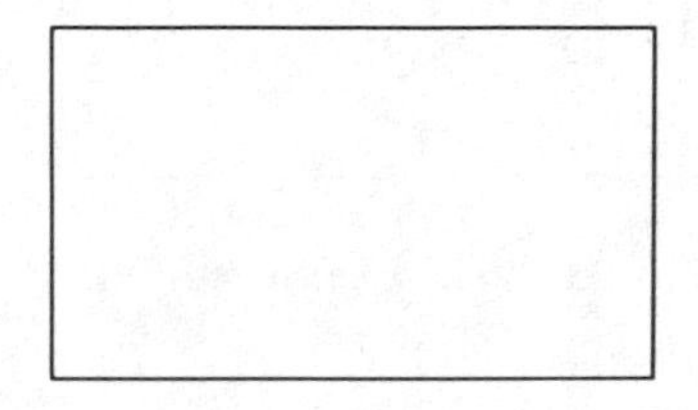

Fig. V1 *Venn diagram, universal set*

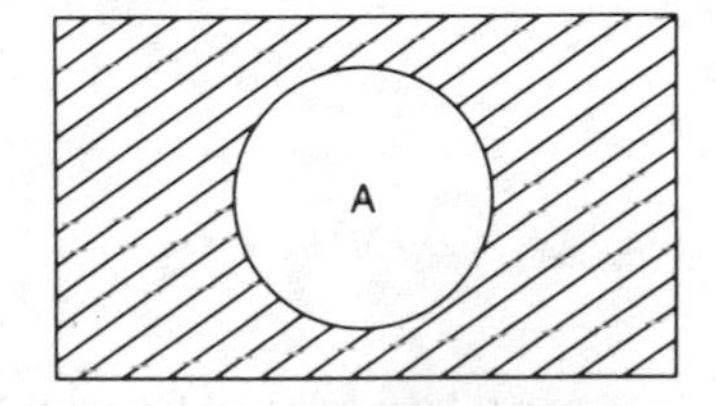

Fig. V2 *Venn diagram, subset and complement*

and the shaded area is the **complement** of set A. The latter represents those elements not in set A, and is denoted by the symbol A′. In Figure V3, the **intersection** of sets A and B is shown; this is written as $A \cap B$. In Figure V4, the **union** of sets A and B is shown. This is written as $A \cup B$.

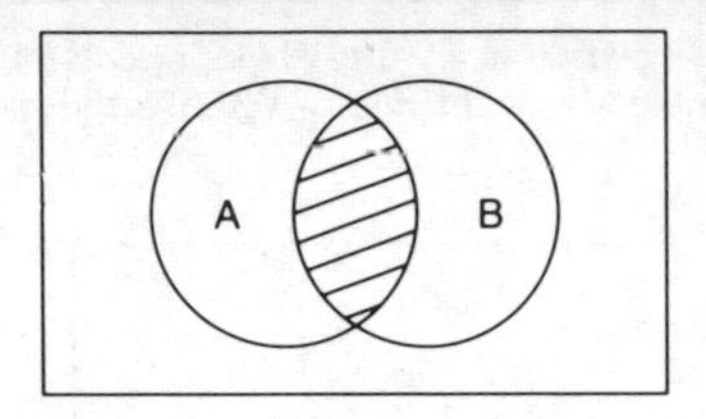

Fig. V3 *Venn diagram, intersection*

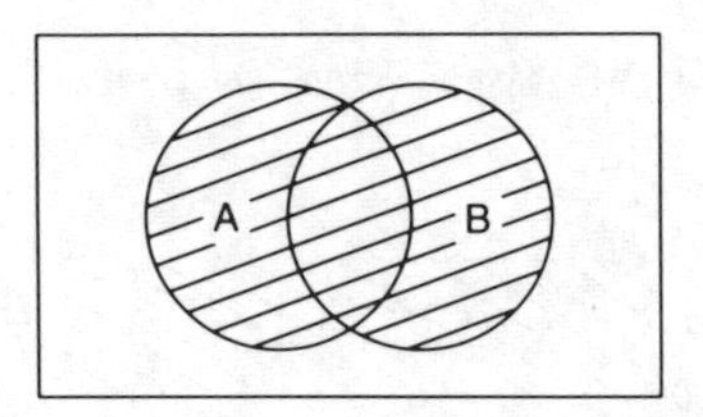

Fig. V4 *Venn diagram, union*

vertex (plural vertices) *1.* **Intersection** of the sides or faces of **polygons** or polyhedra. *2.* Intersection of a **conic section** with its major **axis**.

vertical Direction at right angles to the **horizontal** (the plane of the horizon).

vertices Plural of **vertex**.

visual display unit (VDU) Television-type screen (based on a **cathode-ray tube)** for displaying **alphanumeric** characters or graphics that represent data from a computer or word-processor. Data is entered using an **input device** such as a **keyboard, light-pen** or mouse.

■ **volume** (V) Amount of space occupied by a solid object, or the capacity of a hollow vessel. [4/5/g] The volumes of some common figures are as follows (where l = length, b = breadth, h = height or altitude, r = radius):

cube	l^3	cuboid	lbh
pyramid	$\frac{1}{3}lbh$	sphere	$\frac{4}{3}\pi r^3$
cone	$\frac{1}{3}\pi r^2 h$	cylinder	$\pi r^2 h$

■ **vulgar fraction Fraction** in which both **numerator** and **denominator** are integers (whole numbers). [2/4/k] *See also* **improper fraction**; **proper fraction**.

W

W Abbreviation of **compass** direction West.

wage Money paid to an employee, usually expressed as an hourly, daily or weekly rate. *Compare* **salary**. *See* **gross**; **net**.

wave function Mathematical equation that expresses time and space variation in **amplitude** for a wave system.

weighted mean Average (or **mean**) of a set of numbers which has first been altered to put the numbers into positions of relative importance one to the other (i.e. weighted). *Example*: if a set of numbers X, Y and Z are 25, 30 and 35 respectively, then the mean is 30. If, however, the figures have been weighted 1:2:3 respectively (i.e., 25 is the first, and most important number, 30 is the second, etc.) then the weighted mean will be

$$(1\times 25+2\times 30+3\times 35)/(1+2+3)=(25+60+105)/6$$
$$=31{\cdot}67$$

whole numbers **Natural numbers** that are made up of **integers**. There is no decimal or fractional part to the number.

Winchester disk **Hard disk** and drive that is small enough to use in a **PC**. It was named after the city of Winchester, USA.

word processor *1.* Dedicated microcomputer that is programmed only to prepare text, for data transmission or printing. *2.* Software designed to allow the user to prepare and edit text using a computer.

write head Electromagnetic device that records signals onto a magnetic storage medium (e.g. magnetic tape or disk).

X, Y, Z

X Number 10 in the **Roman numeral** system.

x Symbol used to denote an **unknown** value in a mathematical operation.

***x*-axis** First **axis** of a system of **Cartesian coordinates** x and y.

y Symbol used to denote an **unknown** in a mathematical operation.

yard (yd) **British unit** of length. It is subdivided into 3 feet, or 36 inches = 0·9144 **metres**. One **mile** is equal to 1,760 yards.

***y*-axis** Second **axis** of a system of **Cartesian coordinates** x and y.

yd Abbreviation for **yard**.

zenith In astronomy, the point on the **celestial sphere** directly above an observer's head. *Compare* **nadir**.

zero Value 0, or nought.

zero index Number raised to the power zero. It is always equal to 1, e.g. $x^0 = 1$.

APPENDIX I

Conversion factors

To convert British units to metric units:

	To convert	*To*	*Multiply by*
Length	inches	millimetres	25·4
	inches	centimetres	2·4
	inches	metres	0·245
	feet	centimetres	30·48
	feet	metres	0·3048
	yards	metres	0·9144
	miles	kilometres	1·6094
Area	in^2	cm^2	6·4516
	ft^2	m^2	0·0929
	yd^2	m^2	0·8316
	acres	hectares	0·4047
	acres	km^2	0·00405
Volume	in^3	cc	16·3871
	ft^3	m^3	0·0283
	yd^3	m^3	0·7646
	$miles^3$	km^3	4·1678
Capacity	fl oz	ml	28·5
	pt	ml	568·0
	US pt	ml	473·32
	pt	l	0·568
	US pt	l	0·4733
	gal	l	4·55
	US gal	l	3·785

	To convert	*To*	*Multiply by*
Weight	oz	g	28·3495
	lb	g	453·592
	lb	kg	0·4536
	lb	tonnes	0·000454
	ton	tonnes	1·0161
	US ton	tonnes	0·90719

Metric units to British units

	To convert	*To*	*Multiply by*
Length	mm	in	0·03937
	cm	in	0·3937
	cm	ft	0·03281
	m	in	39·37
	m	ft	3·2808
	m	yd	1·0936
	km	miles	0·6214
Area	cm^2	in^2	0·1552
	m^2	ft^2	10·7636
	m^2	yd^2	1·196
	km^2	$miles^2$	0·3861
	km^2	acres	247·1
	ha	acres	2·471
Volume	cm^3	in^3	0·061
	m^3	ft^3	35·315
	m^3	yd^3	1·308
	km^3	$miles^3$	0·2399

	To convert	*To*	*Multiply by*
Capacity	ml	fl oz	0·0351
	ml	pt	0·00176
	ml	US pt	0·002114
	l	pt	1·760
	l	US pt	2·114
	l	gal	0·2193
	l	US gal	0·2643
Weight	g	oz	0·0352
	g	lb	0·0022
	kg	lb	2·2046
	tonnes	lb	2204·59
	tonnes	ton	0·9842
	tonnes	US (short) ton	1·1023

Temperature

To convert **Fahrenheit** temperatures to Celsius temperatures, subtract 32 and multiply by 5/9.

To convert a **Celsius** temperature to the Fahrenheit scale, multiply by 9/5 and add 32.

Useful formulae

Area

triangle $\frac{1}{2} l \times h$

$\sqrt{(s(s-a)(s-b)(s-c)}$ where

$s=\dfrac{a+b+c}{2}$

$\frac{1}{2} bc \,.\, \sin \mathrm{A}$

parallelogram	$l \times h$
trapezium	$\frac{1}{2}$(sum of parallel sides) $\times h$
rectangle	$l \times h$
rectangular solid (with sides a, b, c)	$2(ab+bc+cd)$
square	l^2
circle	πr^2
sector of a circle	$\frac{1}{2} r^2 \times \theta$(rad)
sphere	$4\pi r^2$
cone (curved surface)	πrs (s = slant height)
cone (total surface)	$\pi rs + \pi r^2$
cylinder (curved surface)	$2\pi rh$
cylinder (total surface)	$2\pi rh + 2\pi r^2$

Perimeter

rectangle	$2l+2h$
square	$4h$
triangle	$a+b+c=2s$ (s = semi-perimeter)
circle	$2\pi r$

Length of arc subtending an angle $\theta = \dfrac{2\pi r\theta}{360}$ or $r\theta$ radians

Volume
(l = length, b = breadth, h = height)

cone	$\frac{1}{3}\pi r^2 h$
cube	l^3
cuboid	lbh
cylinder	$\pi r^2 h$

pyramid	$\frac{1}{3}\, lbh$
sphere	$\frac{4}{3}\, \pi r^3$

Sum of interior angles in a n-sided polygon $= n - 2 \times 180°$

Trigonometrical functions

sin θ	b/c
cos θ	a/c
tan θ	b/a
cosine rule	$a^2 = b^2 + c^2 - 2bc\,.\cos \mathrm{A}$
sine rule	$\frac{\sin \mathrm{A}}{a} = \frac{\sin \mathrm{B}}{b} = \frac{\sin \mathrm{C}}{c}$

Quadratic equations

$$x = \frac{-b \pm (b^2 - 4ac)}{2a}$$

Interest
(I = interest, P = initial sum, R = rate, T = time)

Simple Interest: $\mathrm{I} = \frac{\mathrm{PRT}}{100}$

Compound Interest: $\mathrm{I} = \mathrm{P}(1 + \mathrm{R}/100)^T$

APPENDIX II

Calculator use

The majority of electronic pocket calculators currently fall into three broad types:

(a) simple calculators;
(b) scientific calculators;
(c) graphical calculators.

There are overlaps between the three, for example in calculators dedicated to metric conversion, where the operations are confined to those available on a simple calculator, but where conversion functions are used to allow conversion from, say, **pounds** to **kilograms**, or from °F to °C, and vice-versa.

When using any calculator, however, it is important to remember the priority given in the calculator to different operations and functions. In most simple calculators, % has priority over × and ÷, which in turn have priority over + and −. In most scientific calculators, a bracket operation (. . . . . .) has priority over a function operation (such as log), which has priority over × and ÷. Inattention to these priorities can make the use of a calculator less effective.
Example: keying in

$$36-26\times 93$$

will produce a different result from keying

$$36\times 93-26$$

Again, to calculate 25% of 36 × 132, key 36 × 132 × 25%, = 1188. If 25% 36 × 132 is keyed, an incorrect result will be arrived at.

Most calculators also have a memory function, which can be useful for performing calculations dependent on earlier ones.

Simple calculators
These usually perform the basic $\times$, $\div$, $-$ and $+$ operations, with the functions % and $\sqrt{}$ on an 8-digit display.

Scientific calculators
These usually have a range of inbuilt calculation functions over and above $\times$, $\div$, etc. The standard functions (on 10- or 12-digit displays) include π; **common logarithm**; **Napierian logarithm**; the **trigonometric functions**; **hyperbolic functions**; fraction key; rectangular to polar conversion; **sexagesimal** angular notation; a statistical mode; facilities for calculation in **binary**, **hexadecimal**, **octal** and **decimal** bases, and keys for entering expressions in **brackets**. Such calculators commonly operate with a SHIFT, INV or 2nd Function key (similar to the SHIFT key on a typewriter or PC keyboard) which allows the user to access inverse functions for particular keys, such as $\cos^{-1}$ on the 'cos' key. Some scientific calculators will also perform calculations using **complex numbers**; most have a **program** facility. Some calculators will also retain the various elements of a calculation on screen, in sequence, as they are keyed in, and retrieve the full calculation sequence after = has been keyed. This can be particularly useful if the calculation is a lengthy one, or if the result of it has been an ERROR message.

Graphical calculators
These retain many of the functions of the scientific calculator, but have the added facility to plot and display values of x and

y on the **Cartesian coordinate** system. They generally have extensive hierarchies of memory which can be used to execute programs and, having a larger screen than the average calculator, will display approximately eight lines of single or combined calculation on screen at any one time.

Calculator key functions

Key functions are listed here by symbol, with a brief description of the function, plus examples. The list is intended to be representative, but not exhaustive, and therefore individual keys or functions found on specific calculators may not be included. The manufacturer's handbook should always be studied carefully when using a specific calculator.

Min MR M+ M– These are usually **memory** keys:
Min = enter into memory;
MR = recall from memory;
M+ = add to memory;
M– = subtract from value in memory.
Calculator memory varies in size according to the specific calculator in use. In programmable calculators, memory storage may be accessed by a **STO** key, and subdivided by use of an **ALPHA** function key. In its simplest use, the memory function does away with the need to write down the result of one calculation before going on to second, third, etc. calculations which all depend on one another for their final solution. *Example*: to calculate

3×20
$+16 \div 18$
-3×85

key 3×20 **Min**
$16 \div 18$ **M+**
3×85 **M−**
then **MR**

which supplies the value −194·111′. It's always important when beginning a new calculation using memory to check the memory is clear before you start. Some calculators will retain values in memory when switched off, and may not automatically clear memory when a new value is entered.

± (−) Gives a number keyed subsequently a minus value (NB *not* a **subtraction** operation key). *Example*: calculate -3×-27; key **±** 3 × **±** 27 =, supplies the value 81.

π Inbuilt function for the value of **pi**, usually accessed by an **INV** function key, which allows calculations including π to be made directly. *Example*: find the volume of a **sphere** with **radius** 2·164 m. The formula is $\frac{4}{3}\pi r^3$, therefore key the formula with the values given, i.e. 4 ÷ 3 × π × 2·164 y^x =, which gives the value 42·4483 m^3. It's important here to place the × operators between the values to obtain the correct answer.

EXP The **exponent** key can be used to: *1.* express values as a power of 10. *Example*: raise 2·536 to $\times 10^5$. Key 2·536 **EXP** 5 =, returns the value 253600; lower 1,387·5 by $\times 10^{-3}$-key 1387·5 **EXP** **±** 3 =, returns the value 1·3875. *2.* perform calculations using terms to powers of 10. *Example*: calculate $18 \times 10^5 \div 6 - 3 \times 10^{-1}$. Key 18 **EXP** 5 ÷ (6 − 3 **EXP** **±** 1) =, returns the value 315789·473.

$\sqrt{}$ $\sqrt[x]{}$	The $\sqrt{}$ key returns the **square root** of a number. *Example*: find the square root of 857·16. Key $\sqrt{}$ 857·16=, returns the value 29·27729. The $\sqrt[x]{}$ key will give the xth root of a number. *Example*: find 13568 to the root 6. Key 6 $\sqrt[x]{}$ 13568=, supplies the value 4·88374.
x^2 y^x	The x^2 key will give the **square** of a number; the y^x will multiply a number supplied to the power specified. *Example*: find 18^2+25^5. Key 18 x^2+25 y^x 5=, supplies the value 9765949.
log	This key will give the **common logarithm** of a number; combination with the **INV** key will give the common **antilogarithm**. *Example*: find log 3560÷antilog 3·1120. Key log 3560 ÷ (INV log 3·1120) =, gives the answer 0·00274.
ln	This key will supply the **Napierian logarithm** of a number; with the **INV** key it will give the Napierian antilogarithm.
sin cos tan	These are the **trigonometric function** keys, **sine cosine** and **tangent**; with the INV key, they will give the **inverse** functions **cos**$^{-1}$, **sin**$^{-1}$ and **tan**$^{-1}$. Care should be taken when using these keys to ensure that the correct unit of angle measurement is being used (**degrees**, **radians** or grads). *Example*: find sin 35°; key sin 35 (ensuring degree mode is in use)=0·57357. Find $\tan^{-1}$ 14·30066: key INV tan 14·30066= 85·999=89°. Find cos $\frac{\pi}{6}$ radians: select radian mode, key cos $\pi \div 6$=0·866025.

Conversion: degrees; decimal degrees; radians Most scientific calculators allow the user to enter angles as degrees, minutes and seconds (**DMS** or °, ′, ″ key), as decimal degrees, or as radians, and to convert from one into the other. *Example*: enter 35° 20′ 15″ as degrees, minutes and seconds. Key 32 DMS 20 DMS 15 DMS. Convert to decimal degrees: key INV DMS, supplies 32·3375. Convert to radians: key INV **DRG**, supplies 0·56439 radians.

HYP When used in combination with the sin, cos and tan keys, HYP will return the **hyperbolic functions sinh, cosh** and **tanh**; used with the INV key it will give $\sinh^{-1}$, $\cosh^{-1}$ and $\tanh^{-1}$. *Example*: find sinh when $x=2{\cdot}235$: Key HYP sin 2·23 =, gives 4·59616. Find $\cosh^{-1}$ 5·667: key INV HYP cos 5·667 =, which gives 2·41993.

$a\frac{b}{c}$ This key allows calculations to be performed using **vulgar fractions**, and conversion from vulgar fractions to decimal fractions. *Example*: calculate $23\frac{5}{6}\times\frac{3}{18}\div 6\frac{1}{3}$. Key 23a $\frac{b}{c}$ 5a $\frac{b}{c}$ 6 × 3a $\frac{b}{c}$ 18 ÷ (6a $\frac{b}{c}$ 1a $\frac{b}{c}$ 3) =, which gives $\frac{143}{228}$.

Pressing $a\frac{b}{c}$ when a vulgar fraction is displayed converts the expression to a decimal fraction: $\frac{143}{228}$ a $\frac{b}{c}$ returns the value 0·62719.

(... ...) Parentheses (**bracket**) keys allow calculations to be performed which include expressions enclosed within brackets. *Example*: calculate $28+\pi^2(3\div 18)$. Key 28 + πx^2 × (3 ÷ 18) = which gives 29·64493. Note that the × operator between π^2 and $(3\div 18)$ must be included to give a correct solution.

$\frac{1}{x}$ or x^{-1} These keys find the **reciprocal** of any value (some calculators use the first symbol, whilst others use the second). Apart from calculating ordinary reciprocals, this key can be used to find such values as the **trigonometrical functions cosecant**; **cotangent** and **secant**.

Example: find sec 35°. Key cos 35° $\frac{1}{x}$, which gives the value 1·220779. Calculators with a x^{-1} key require the calculation to be keyed as: cos 35° $= x^{-1} =$ to give the correct solution.

***x*!** This key calculates the **factorial** of a number, and can therefore be used to calculate values such as the **combination** of objects in a set. *Example*: calculate the combination of 6 objects from a set of 8. Key $8x!\div((6x!\times(8-6)x!)$, which gives 28.

DEC BIN OCT HEX Most scientific calculators allow calculation in **decimal** (DEC), **binary** (BIN), **octal** (OCT) and **hexadecimal** (HEX) number bases, and conversion from one to the other. *Example*: add binary 11001 + 100100 and convert to decimal. Engage BIN mode, key 11001 + 100100 INV DEC, which gives the value 61.

^nC_r This key will calculate the **combination** of objects in a set. *Example*: how many combinations of 8 objects can be obtained from a set of 12 objects? Key 12 ^nC_r 8, which supplies the value 495.

^nP_r This key will calculate the **permutations** possible in a given set. *Example*: how many six figure number permutations can be found in the set of 8 numbers 1–8? Key 8 ^nP_r 6, which supplies the value 20160.

Rec/Pol This function will convert **polar coordinates** supplied as a length r and an angle θ to rectangular **(Cartesian)** coordinates, i.e. values (x, y), and vice-versa. *Example*: find Cartesian coordinates for a point P with $r=6$ and $\theta=52°$. Key 6 Rec 52 =, which supplies the value 3·6939, therefore $x=$ 3·69. Key RCL plus the appropriate y function key, which gives 4·7280, therefore $y=4·73$. Find θ and r for Cartesian coordinates 3·5 and 6. Key 3·5 Pol 6 =, which gives 6·9462, therefore $r=6·94$. Key RCL plus the appropriate θ function key, which supplies the value 59·7435, therefore $\theta=60°$. Note that most calculators will perform the conversions using the same basic steps, but the precise key sequence and the names of the relevant function keys for any specific calculator should be checked in the specific manufacturer's guide.

i Some calculators will perform calculations using **complex numbers**: check that this function is available before attempting to calculate using $\sqrt{-1}$.

programs Most scientific calculators offer the facility to a greater or lesser extent to create and run programs. Programming usually requires the facility to enter terms as letters of the alpahabet (A, B, C etc.): an ALPHA key is the usual key for performing this function. Programs can range in complexity from, for example, one which will find the volume of a cone to one which will perform **prime number** analysis.

graphical Graphical calculators include **plot**, **graph**, **line** and **trace** keys which allow the user to create graphs on the Cartesian coordinate system and to use such graphs in a wide variety of ways. The range of the graph (i.e., the scale and maximum values for x and y coordinates) can be altered at will. A graph for a given value of y can be created, and the changing values of the x coordinate found, for example. Graphs can be enlarged by specified factors, and the graph-creation facility can be linked to statistical data to create bar graphs and to find a normal distribution curve for a set of data.

Statistical functions Most scientific calculators will perform a range of calculations on input statistical data. The key functions normally available are:

$\bar{x}$ **arithmetic mean** of samples;

Σx sum of samples;

Σx^2 sum of the square of samples;

σn **standard deviation** of samples;

σn^{-1} population standard deviation of samples. The usual procedure is to put the calculator into statistical mode, and then enter each item of data, keying a **DATA** (D) entry key after each item. *Example*: calculate the mean and standard deviation of the distances travelled by six people attending a meeting together, where they travelled 80, 312, 94, 67, 40, and 82 kilometres respectively. Engage statistical mode; key 80 D 312 D 94 D 67 D 40 D 82 D. Then key $\bar{x}$, which gives 112·5; key σn, which gives 90·78. The six people therefore travelled an average of 112·5 km each, and the standard deviation of the distances is 90·78 km. Note that the key press sequences for engaging statistical mode, entering data, and finding the different values will vary according to the specific calculator being used, and the manufacturer's guide should be checked for these points.

APPENDIX III

Brief biographical details of people who made important contributions to, or whose names are given to, concepts referred to in this Mini Dictionary are listed in this Appendix, cross-referred in **bold** to relevant entries in the Mini Dictionary.

Argand, Jean Robert (1768–1822) Swiss mathematician. Very little is known about Argand's life, save that he lived in Paris in adult life. In 1806 he published a book which sets out the graphical method of representing **complex numbers** which now bears his name. Other mathematicians had arrived at the same idea at roughly the same time. Argand, however, had done so quite independently, and was the first person to publish it.

Babbage, Charles (1792–1871) English mathematician who in the early 19th century founded a society for for the development of approaches to **calculus**, and was professor of mathematics at Cambridge from 1828–39. Babbage is best known for his researches into ways of creating mechanical **calculators**. The basic designs which he created were sound; the engineering techniques to build them had not been perfected, however, and he died without seeing any of his projects realised.

Bayes, Thomas (1702–61) English mathematician and Nonconformist minister who developed a mathematical form of stating the **probability** that a given event is the result of another event, the probability of which is already known. Bayes's ideas were not in fact published until after his death. *See* **Bayes's theorem**.

Bernoulli, Daniel (1700–82) Member of a famous Swiss family of scientists and mathematicians which began with Jakob Bernoulli (1654–1705) and his brother Johann (1667–1748), both of whom worked extensively on **calculus**. Daniel was Johann's son; he trained in medicine, and was professor successively of botany, anatomy and natural philosophy. His most important work was a publication on the dynamics of fluids. *See* **Bernoulli's theorem**.

Bode, Johann (1747–1826) German astronomer who became director of the Berlin observatory in 1786, founded an important German astronomical yearbook, and published extensively on astronomy. *See* **Bode's law**.

Boole, George (1815–64) Largely self-taught English mathematician who, after 18 years as a teacher and headmaster, was appointed professor of mathematics at Queen's College, Cork. Boole's great achievement was his creation of a method of applying the rules of **algebra** to statements in **logic**; a fundamental development of the concept of **sets** was integral to this work. *See* **Boolean algebra**.

Cantor, Georg (1845–1918) German mathematician who taught as professor of mathematics at Halle University. His life was passed in relative obscurity, and it was not until after his death that his work on infinite **sets** received attention. Since then, however, his ideas have acquired great significance in the approaches used in modern mathematics.

Celsius, Anders (1701–44) Swedish astronomer, professor of astronomy at Uppsala University, who published observations on the aurora borealis, and took part in an expedition to measure the arc of the **meridian**. He is best known for his invention of the **centigrade** thermometer, presented in a paper to the Swedish Academy in 1742. *See* **Celsius scale**.

Descartes, René (1596–1650) French philosopher, scientist and mathematician, Descartes is one of the great figures of European ideas. He was born in France, but spent most of his life elsewhere, most notably in Holland, where he lived for almost 20 years. Descartes thought and wrote about a very wide range of subjects. He is, in many ways, the epitome of the historical movement associated with the term 'Enlightenment', which aimed towards unifying in a single consistent system all human knowledge. He contributed very significantly to European philosophy, but also to physiology, optics, physics, astronomy and biology. His great contributions to mathematics lay in the founding of **coordinate geometry**, the classification of curves and the theory of **equations**. *See* **Cartesian coordinates**; *see also* **Fermat**.

Einstein, Albert (1879–1955) A theoretical physicist, Einstein was born in Germany, but rejected his citizenship and was successively a Swiss and US citizen. His *Special Theory of Relativity*, published in 1905, and *General Theory of Relativity* (1916) completely reshaped the view of the physical world put forward in the writings of **Newton**. In the Special Theory, time ceased to be a constant, and was presented as relative to the speed of an observer. In the General Theory, space was presented as curved by the mass of bodies; light does not move in a perfectly straight line, but will follow a curve. *See* **Newtonian mechanics**; **relativity**.

Eratosthenes (*c.* 275–192 BC) Greek scientist and mathematician, friend of Archimedes, he succeeded in calculating the circumference of the Earth to within a very small degree of error, and is believed to have written a catalogue of over 500 stars. He also constructed a calendar using the concept of a **leap year**, and created the method for finding **prime numbers** known as the **sieve**, or Eratosthenes' sieve.

Euclid (*fl.* 100 BC) Greek mathematician, and the author of the longest-lived textbook in history, the *Elements*. Very little is known of his life, apart from the fact that he founded a school at Alexandria, one of the great centres of learning in classical times. The *Elements* is essentially a summary for students of the discoveries of the Greek mathematicians (such as **Pythagoras**), and is divided into 13 books. Books I-III discuss concepts and problems in **geometry** such as line, point, **congruence**, **area**, the circle, **chords** and **tangents**. Book V deals with **proportion**, Books VII-IX discuss questions in **arithmetic**, Book XI discusses the geometry of **solids**, and Book XIII discusses the five **regular polyhedra**. The *Elements* was not the first attempt to gather together the then basics of mathematical knowledge. It was, however, the most rigorously written and clearly organized, and was acknowledged in classical times as the absolute foundation of mathematics. It was translated in succeeding centuries into Arabic and Latin, and remained throughout the Middle Ages and into the modern period the prime source for mathematical teaching and discussion in Europe. Other works by Euclid have survived, such as a treatise on optics and a work on the geometry of the sphere. *See* **Euclidean geometry**.

Euler, Leonhard (1707–83) Swiss, and a pupil of Johann **Bernoulli**, Euler is considered, along with **Descartes** and **Fermat** as one of the founders of modern mathematics. His adult life was spent alternately in Russia and Germany where he occupied important positions in St Petersburg and Berlin. His interests were broad, and he contributed to a whole range of ideas in astronomy and physics as well as in mathematics, but his fame rests on his work on **integral calculus** and **differential calculus**. Euler was also the first person to systematize the use of many of the symbols used in modern mathematics, such as

π, e, Σ, **log** x, i and $f(x)$. *See* **complex number**; **Euler's formula**; **function**; **imaginary number**; **Königsberg bridge problem**.

Fahrenheit, Gabriel (1686–1736) German physicist who invented the mercury-in-glass thermometer, and the temperature scale named after him. Fahrenheit chose at random to divide the distance between the height of the mercury column in melting ice and boiling water by 180°, thus giving the 32 °F and 212 °F measures for these temperatures. The Fahrenheit scale, although now largely superseded by the **Celsius scale**, has been favoured against the latter because it allows for finer measurement.

Farey, John (1766–1826) English geologist and mathematician, inventor of the **Farey sequence**, whose son of the same name achieved prominence as a civil enginer. John Farey senior established himself as a consultant surveyor and geologist, and in the course of his career travelled widely throughout England taking geological samples.

Fermat, Pierre de (1601–65) French mathematician and contemporary of **Descartes**, Fermat made significant contributions to several branches of mathematics. He invented the basis of **coordinate geometry** before Descartes but, since he did not publish his ideas, Descartes is generally credited with the achievement. Fermat also arrived at the fundamental principles of **calculus**, invented a number of theorems for calculating **prime numbers** and, with **Pascal**, founded the modern mathematical theory of **probability**.

Fibonacci, Leonardo (Leonardo of Pisa) (*c.* 1180–*c.* 1250) Mediaeval Italian mathematician who in his youth lived in North Africa, and later travelled extensively throughout the Mediterranean. He published several treatises, the most famous of which is his *Book of the Calculator* in which (among other things) he proposed the adoption of the **Arabic number system**, including the use of 0, until that date little known in Europe. The book discusses methods and problems in **algebra**, the use of **fractions**, presents tables for **multiplication** and **prime numbers** and, in its final part sets out various mathematical games, among which appears the concept for which the author is now best known, the **Fibonacci series**.

Hero (Heron) of Alexandria (*fl. c.* 62 AD) Greek scientist and mathematician who wrote treatises on, for example, different kinds of machinery, as well as on **geometry** and measurement. We know when he lived because he refers in one of his writings to a lunar eclipse at Alexandria which can be dated accurately. His proof for his formula for finding the area of a triangle appears in the same text. *See* **Hero's formula**.

Hooke, Robert (1635–1703) English physicist and contemporary of **Newton**, Hooke produced work on the theory of light, on astronomy, and on gravitation. He was also an inventor, and is credited in this field with the invention of the wheel barometer, and the use of spiral springs in watches. *See* **Hooke's law**.

Kepler, Johannes (1571–1630) German astronomer and mathematician. Kepler first taught as a professor of mathematics, and it was not until he moved to Prague in 1600 and succeeded to the post of court astronomer there that he really began to develop his ideas about the motion of the planets. The Ptolemaic (or Earth-centred) theory of planetary motion was still very strongly held throughout Europe at that time. Although he knew the ideas of Copernicus, and was a friend of Galileo (both of whom evolved a Sun-centred theory), Kepler supported the Ptolemaic view, but was concerned to discover a consistent pattern in the theory. His first step was to propose that the planets moved in ellipses, and from this followed the concepts we now know as **Kepler's laws**. Although Kepler did not put the idea forward, his laws were very important in the proof that the the planets do, in fact, move around the Sun. He is also credited with being among those mathematicians who contributed to the use of **logarithms**, and to the developent of **calculus**.

Leibniz, Gottfried Wilhelm (1646–1716) German philosopher, mathematician and diplomat, Leibniz can be seen as belonging in the 'universal genius' niche occupied by such other great figures as **Descartes**. He published essays on law, philosophy and mathematics before the age of 21, pursued a career as a diplomat and political adviser to several European courts, made a major contribution to European logic and philosophy and, along with **Newton**, was one of the founders of **calculus**. Among his many other accomplishments was the invention of a mechanical **calculator** which improved on the design for such a machine put forward by **Pascal**.

Lissajous, Jules Antoine (1822–80) French physicist, Lissajous taught physics, and was rector of various provincial French academies. He evolved an ingenious method for displaying a visual image which demonstrated the **amplitude** and frequency of sound waves. It involved the use of mirrors placed at right angles to each other. *See* **Lissajous figure**.

Lorentz, Hendrik Antoon (1853–1928) Dutch physicist who received the Nobel Prize for physics in 1902, and whose work helped to lay the basis for **Einstein's** *Special Theory of Relativity*. *See* **Lorentz transformation**.

Mercator, Gerardus (1512–94) Flemish surveyor, instrument maker and cartographer, he was the first mapmaker to use the projection named after him on a nautical map which could be used to set a **compass** course. *See* **Mercator's projection**.

Möbius, August Ferdinand (1790–1868) German astronomer and mathematician who contributed to the development of **analytic geometry**, projective geometry and topology. His work on the last included a study of one-sided surfaces, among them the **Möbius strip**.

Napier, John (1550–1617) Scottish mathematician who contributed to the theory of spherical geometry, invented a calculation system based on the manipulation of small rods known as 'Napier's bones', and invented the system of describing a number as a **base** to a given **power** known as **logarithms**. Napier was also a vehement debater on religious questions, and a manuscript describing various military machines bears his signature.

Newton Sir Isaac (1642–1727) English scientist and mathematician, his theories of gravity and the composition of light transformed Western scientific thought. As a mathematician, he independently arrived at the methods of integral and diferential **calculus**: a dispute with **Leibniz** over which of them should receive the credit for the discovery helped to embitter the later lives of both men. *See* **newton**; **Newtonian mechanics**; **Newtonian method**.

Pascal, Blaise (1623–62) French mathematician, his short life was dogged by illness and religious controversy. Pascal contributed to projective geometry through his work on **conic sections**, made some important observations about the **cycloid** and, in collaboration with **Fermat** established the combinatorial approach to calculating **probability**. He also invented a mechanical **calculator**. *See* **combination**; **Pascal's triangle**.

Pythagoras (*fl. c.* 530 BC) Greek philosopher and mathematician who founded a school of philosophy and mathematics in southern Italy from which grew a system of beliefs about the world which we now term 'Pythagoreanism'. The Pythagoreans believed, for example, that the soul moved from one body to another after death. Mathematics, particularly number theory and the connections between mathematics and music, were also central to the school's preoccupations. Much study was given, for example, to the properties of **square** numbers. Geometry was also studied: the discovery that the **interior angles** of a triangle = 180° is credited to the Pythagoreans.

Simpson, Thomas (1710–61) Largely self-taught English mathematician who, from being a weaver who taught mathematics part-time, grew to be a Fellow of the Royal Society. He published papers on **calculus**, **probability**, **algebra**, **geometry** and trigonometry. In addition to devising **Simpson's rule**, he also discovered a method for calculating the volume of a *prismoid*.

Taylor, Brook (1685–1731) English mathematician who carried **Newton's** work on **calculus** forward, applying it for the first time to the vibration of a string, and creating the series to describe a **function** which bears his name.